Biodiversity and Environment

BIODIVERSITY AND ENVIRONMENT

By

Dr. M. Lakshmi Narasaiah

M.A., Ph.D.

Professor of Economics

Sri Krishnadevaraya University Post-graduate Centre

Kurnool – 518 002

Andhra Pradesh

(India)

DPH

DISCOVERY PUBLISHING HOUSE PVT. LTD.

NEW DELHI-110 002

First Published – 2004

Reprinted – 2026

ISBN: 978-81-7141-758-2

Biodiversity and Environment

Published by:

DISCOVERY PUBLISHING HOUSE

4383/4B, Ansari Road, Darya Ganj

New Delhi-110 002 (India)

Phone: +91-11-23279245; 23253475; 43596065

Mobile: +91 9811179893 / +91 9871656464

E-mail: discoverybooksindia@gmail.com

orderdphbooks@gmail.com

namitwasan9@gmail.com

web: www.discoverypublishinggroup.com

Printed at:

Infinity Imaging Systems

Delhi

Preface

As human population has surged this century, the populations of numerous other species have tumbled, many to the point of extinction. Indeed, we live amid the greatest extinction of plant and animal life since the dinosaurs disappeared some 65 million years ago, with species losses at 100 to 1,000 times the natural rate. But humans are not just witnesses to a rare historic event, we are actually its cause. The leading sources of today's species loss, habitat alteration, invasions by exotic species, pollution, and overhunting are all a function of human activities.

Human activities have pushed the percentage of mammals, amphibians, land fish that are in "immediate danger" of extinction into double digits. The principal cause of species extinction is habitat loss—the result of encroachment by humans for settlements, for agriculture, or to claim resources such as timber. A particularly productive but vulnerable habitat is found in coastal areas, home to 60 per cent of the world's population. Coastal wetlands nurture two-thirds of all commercially caught fish, for example. And coral reefs have the second highest concentration of biodiversity in the world, after tropical rainforests. But human encroachment and pollution are degrading these areas: roughly half of the world's salt marshes and mangrove swamps have been eliminated or radically altered, and two-thirds of the world's coral reefs have been degraded, 10 per cent of them " beyond recognition". As coastal migration continues—coastal dwellers could account for 75 per cent of world population within 30 years—the pressures on these productive habitats will likely increase".

Habitat loss tends to accelerate with an increase in a country's population density. This is bad news for the world's biodiversity hotspots-species-rich ecosystems at greatest risk of destruction. Twenty-four of these hotspots, containing half of the planet's species, have been identified globally. Some of the most important hotspot countries will reach population densities that have been linked with very high rates of hibitat loss. Five of the six mcst biologically rich countries could see more than two-thirds of their original habitat destroyed by 2050 if this historical relationship holds.

Related to loss of habitat is the growing incidence of plant, animal, insect, and microbial invasions of ecosystems worldwide as human interchange increases. These "exotic species" sometimes dominate local ecosystems, eliminating native species and reducing overall diversity. Exotics are implicated in 68 per cent of all fish extinctions in the United States this century, for example. Growth in human travel and commerce explains many accidental invasions by exotics, but foreign species are also deliberately introduced into farms, plantation forests, and aquaculture systems. Although only 1 per cent of exotics cause widespread damage, exotic species are the second leading cause, after habitat destruction, of species loss worldwide.

Dr. M. Lakshmi Narasaiah

Contents

1
Biodiversity

As human population has surged this century, the populations of numerous other species have tumbled, many to the point of extinction. Indeed, we live amid the greatest extinction of plant and animal life since the dinosaurs disappeared some 65 million years ago, with species losses at 100 to 1,000 times the natural rate. But humans are not just witnesses to a rare historic event, we are actually its cause. The leading sources of today's species loss, habitat alteration, invasions by exotic species, pollution, and overhunting are all a function of human activities.

Human activities have pushed the percentage of mammals, amphibians, land fish that are in "immediate danger" of extinction into double digits. The principal cause of species extinction is habitat loss—the result of encroachment by humans for settlements, for agriculture, or to claim resources such as timber. A particularly productive but vulnerable habitat is found in coastal areas, home to 60 per cent of the world's population. Coastal wetlands nurture two-thirds of all commercially caught fish, for example. And coral reefs have the second highest concentration of biodiversity in the world, after tropical rainforests. But human encroachment and pollution are degrading these areas: roughly half of the world's salt marshes and mangrove swamps have been eliminated or radically altered, and two-thirds of the world's coral reefs have been degraded, 10 per cent of them "beyond recognition". As coastal migration continues—coastal dwellers could account for 75 per cent of world population within 30 years—the pressures on these productive habitats will likely increase".

Habitat loss tends to accelerate with an increase in a country's population density. This is bad news for the world's biodiversity hotspots-species-rich ecosystems at greatest risk of destruction. Twenty-four of these hotspots, containing half of the planet's species, have been identified globally. Some of the most important hotspot countries will reach population densities that have been linked with very high rates of hibitat loss. Five of the six most biologically rich countries could see more than two-thirds of their original habitat destroyed by 2050 if this historical relationship holds.

Related to loss of halbitat is the growing incidence of plant, animal, insect, and microbial invasions of ecosystems worldwide as human interchange increases. These "exotic species" sometimes dominate local ecosystems, eliminating native species and reducing overall diversity. Exotics are implicated in 68 per cent of all fish extinctions in the United States this century, for example. Growth in human travel and commerce explains many accidental invasions by exotics, but foreign species are also deliberately introduced into farms, plantation forests, and aquaculture systems. Although only 1 per cent of exotics cause widespread damage, exotic species are the second leading cause, after habitat destruction, of species loss worldwide.

Other, often difusse effects of expanded human activities also disrupt ecosystems. Nitrogen, for example, is now made available to plants at more than twice the preindustrial rate as a result of fertilizer production, cultivation of nitrogen-fixing crops, and the burning of fossil fuels. This overfertilization of the Earth favours some species at the expense of others, leading to a reduction in diversity and resiliency of land and aquatic ecosystems.

Likewise, greenhouse gas emissions could distrupt ecosystems on a vast scale. As with nitrogen, increased levels of atmospheric carbon may favour some species over others: annuals over perennials, for example, or deciduous trees over evergreens. To the extent that greenhouse gases induce changes in global climate, many species may be at risk as habitats shift or shrink, and as some life forms, such as insects or animals, adapt and migrate more quickly than others, such as plants. And as sea levels rise with a change in climate, ecosystems such as coastal wetlands could be destroyed.

2

Saving the Planet

Imperialism in Green Garb?

Developing countries feel that protecting the world's resources is just another way for rich nations to retain the upper hand in the international trade game. For nearly a decade, international efforts to address global environmental concerns have been frustrated by a deep rift in perceptions between rich and poor countries. Economists and environmentalists in developing nations argue that the North almost exclusively drives the agenda for environmental negotiations. Under the pretext of saving the planet, they say, the industrialized world is wielding a new brand of dominance, "ecoimperialism".

Developing countries like India and China continue to resist global environmental protocols, like the 1989 Montreal Accord to cut the production of CFC gases (used for example in refrigerators) by 50 per cent, or the Clean Development Mechanism (CDM), part of the climate change negotiations initiated under the 1997 Kyoto Protocol.

The spectre of imperialism is likely to vitiate the next round of climate change talks in Bonn (Germany) when policymakers finalize the terms on which the CDM will be implemented. Negotiated by industrialized countries to gain some flexibility in meeting the emission reduction targets pledged in Kyoto, few issues in recent environmental diplomacy are proving contentious.

Critics say the mechanism is the latest in a string of attempts to dominate poor countries, which are being virtually "bribed" so that rich nations can continue business as usual. By financing forestry schemes and other energy-efficient projects, industrialized countries could exploit the mechanism to avoid reducing their own greenhouse gases. Environmentalists fear this could turn the Amazon and other primeval forests into "carbon sinks" to absorb pollution, but with side effects which disregard the development needs of southern countries.

Lopsided Negotiations

The northern bias continues to dominate discussion of the global climatic crisis. The threat to the atmospheric commons has been building over centuries, mainly because of industrial activity in the North. Yet discussions seem to focus more on developing countries; the North refuses to assume extra responsibility for cleaning up the atmosphere. No wonder the Third World cries foul when it is asked to share the costs.

The 1989 Basel Convention, for instance, imposed restrictions on trade in scrap metals and recyclable materials, claiming they were hazardous to the environment. Economists say it prohibits poor countries from competing in the lucrative world market for computer parts, scrap metal, and recyclable products.

Other examples of trade restrictions are cited. In the early 1990s, Malaysia and Indonesia fought to overturn and ecolabelling law introduced by Austria ostensibly to safeguard the Asian rain forests. Austria refused to import timber that was not from sustainable managed forests, but no such curbs existed for wood for temperate areas. The protectionist flavour of the measure was overt, and Austria eventually revoked it.

In other trade-environment disputes over the last decade, the United States has been accused of protectionism in banning the import of Mexican tuna because dolphins were getting ensnared and killed in nets meant for the fish. Shrimp from India, Pakistan, Thailand and Malaysia, which paid no heed to sea-turtle protection, were similarly banned in 1996. The sanctions may have been motivated by a desire to protect dolphins and turtles, but the poorer countries claimed that they were a pretext for suppressing competition in the global fish market.

A Green Agenda to Stop Growth?

Rules restricting trade through the Basel Convention or attempts to ban genetically modified foods are designed to exclude poorer countries from world markets. Other voices in the South, however argue that environmental controls like the CDM are not all bad. Developing countries, say experts, will receive $5 to $17 billion to fund climate-friendly technologies. The CDM gives us an opportunity to invest in projects that promote sustainable development. If incidentally they also reduce emissions, we shouldn't quarrel with the fact.

3

Forests

Global losses of forest area have marched in step with population growth for much of human history. The two trends rose slowly for millennia, turned upward in recent centuries, and accelerated sharply after 1900. Indeed, 75 per cent of the historical growth in global population and an estimated 75 per cent of the loss in global forested area have occurred in the twentieth century. The correlation makes sense, given the additional need for farmland, pastureland, and forest products as human numbers expand. But sing 1950, the advent of mass consumption of forest products has quickened the pace of deforestation.

In some cases, population pressure is still closely linked with deforestation. In Latin America, for example, ranching is the single largest cause of deforestation. Because most meat produced in Latin America is consumed there, and because meat consumption per person has been largely unchanged for several decades, it is likely that expanding population is the principal reason for ranching-related deforestation. In addition, analysts at the World Resources Institute estimate that overgrazing and overcollection of firewood—which are often a function of a growing population—are degrading some 14 per cent of the world's threatened frontier forests (large areas of virgin forests). In fact, a U.N. Food and Agriculture Organisation study showed a one-to-one correlation between population growth and fuelwood consumption in 16 Asian countries between 1961 and 1994.

On the other hand, deforestation created by the demand for forest products tracks more closely with rising per capita consumption in recent decades, Global use of paper and paperboard per person, for example, has doubled (or nearly tripled) since 1961, and most of the increase has come in wealthy countries with low or even stable levels of population growth. Europe, Japan, and North America, with 16 per cent of global population, consume 63 per cent of the world's paper and paperboard and nearly half its industrial wood.

Although consumption and population growth have operated somewhat independently in the late twentieth century, the two forces could coincide in the developing world in coming decades, with substantial consequences for forests. Developing-country paper consumption is less than one-tenth the level found in industrial nations, suggesting that large increases in consumption are likely as these nations prosper. (It also suggests that greater economy is needed in industrial countries). With 80 per cent of the world's people, and as home to all the increase in population in coming decades, even modest growth in per capita paper and wood consumption in developing countries could place substantial pressure on forests. If paper were used by the entire world in 2050 at today's industrial-nation rates, paper production would need to jump more than eightfold over 1996 levels.

This projected growth is unsustainable, given that global use of forest products is already near or beyond the limits of sustainable use. Using data on sustainable forest yields, and assuming that virgin forests are left intact, researchers at Friends of the Earth UK have determined that production of forest products for the world is 25 per cent beyond the most restrictive estimates for sustainable consumption. (Many forests, of course, are already logged well beyond sustainable levels). The most optimistic assessment would allow for a further 35 per cent growth in consumption. Even that spells trouble, however, given a projected global population increase of some 54 per cent over the next half-century, and given the likely increase in consumption from rising prosperity. Lower consumption of forest products and increased recycling in industrial countries can make room for a more prosperous developing world to enjoy the products of the world's forests, but the task will be made easier if population growth everywhere is stabilized sooner rather than later.

If population and consumption eat into the world's forests, the resulting loss of forest services reduces, in turn, a country's capacity to support its population. Forests provide habitat to a diverse selection of wildlife; tropical forests, for example, are home to more than 50 per cent of the world's species. And as storehouses of carbon, forests are key to regulating climate. Deforestation leads to huge releases of carbon: an estimated one quarter of the world's carbon emissions come from forest clearing. Loss of these macroservices undermines the stability and resiliency of the global environment on which economies—and populations—depend. In addition, forests provide services vital to a local population, such as control of erosion, steady provision of water across rainy and dry seasons, and regulation of rainfall. Taken together, the loss of these services due to deforestation can upset local economies and subject local populations to economic instability.

4

Forests—The Earth's Lungs

The world's forest cover is shrinking. Over the past 50 years nearly half of the world's original forest cover has been lost—some 3 billion hectares. Each year another 16 million hectares of virgin forest are cut, bulldozed, or burned.

Between 1980 and 1995 the world lost some 180 million hectares of forest—an area the size of Indonesia. While developed countries had a net increase of 20 million hectares due to reforestation, this gain was more than offset by a net decrease of 200 million hectares in the developing world.

Forests have many functions of value both to humanity and to nature itself. Take away the trees, and the intricately linked ecosystem unravels. Forests absorb carbon dioxide and produce oxygen, anchor soils, regulate the water cycle, protect against erosion, and provide a habitat for millions of species.

Forest products are essential to the world economy, worth about US$400 billion annually in timber, pulp, paper, and fuel wood. Forest products other than wood, such as medicines, vegetables, and fruits, provide another US$20 billion and are growing in importance.

Healthy forests boost food production. Trees soak up and store water from season to season, slowly releasing moisture during dry periods. Without tree cover, water runs off faster during the tropical rainy season, carrying way valuable topsoil. A World Bank study found that the rate of soil loss was 10 times higher on forest lands where slash-and-burn shifting cultivation was practised than in undisturbed forests. One reason that agricultural yields have fallen in sub-Sahran Africa is that vast amounts of forest cover have disappeared, hastening soil erosion and loss of soil nutrients.

Forest cover regulates climate, while destruction of forests contributes to global warming. Whereas living trees soak up and store corbon dioxide from the atmosphere trees that are cut down and burned release carbon into the atmosphere. In the last decade tropical deforestation has released large amounts of stored carbon—accounting for roughly one-quarter of the carbon dioxide emissions to the atmosphere due to human activity.

Pressures on Forests

Current demand for forest products may exceed the limits of sustainable consumption by 25 per cent. The developed world accounts for most of the demand for forest products. With just 16 per cent of the world's population, North America, Europe, and Japan consume two-thirds of the world's paper and paperboard and half its industrial wood. Demand for industrial wood products also has risen in developing countries, however, along with demand for fuel wood, the main energy source for many rural communities.

Throughout the 1990s many developing countries with rapid population growth had high rates of deforestation. Forest land was converted to agricultural use, and trees cut to provide housing and wood for fuel. Moreover developing countries stepped up exports of forest products to meet the rising demand from developed countries.

The amount of forest area per capita fell by half between 1960 and 1995—reflecting both population growth and the disappearance of forest cover. In 1995 close to 1.7 billion people lived in countries with less than one-tenth of a hectare of forest cover per capita (83). By 2025, an estimated 4.6 billion people will live in such countries.

What Can be Done?

As population grows and per capita consumption of forest products increases, countries must do more to manage forest resources on a sustainable basis. The following developments offer encouragement.

Technological Improvements

Technological improvements including use of recycled paper and paperboard, have substantially reduced the amount of pulp needed to produce paper. In 1970 paper and paperboard consisted of 80 per cent wood pulp. By 1997 more efficient production processes had reduced that figure to 56 per cent. As a direct result, the production of pulp for paper is expected to grow by just over 1 per cent a year over the next decade, about half the growth rate in the 1980s.

Forest Products Certification

Adopting a system that identifies forest products that come from sustainable managed forests could support toward sustainability. As of 1998, about 10 million hectares of forest lands have been certified. Over 90 per cent of the certified area is in northern, temperate forests, mostly in Europe and north America. Close to 60 per cent of the entire certified area is in just two countries—Sweden and Poland—reflecting education and awareness campaigns in those countries. In tropical forest, where most of the destruction is taking place today, only tiny areas have been certified as providing sustainable yield.

Intergovernmental Responses

In 1995 the Intergovernmental Panel on Forests (IPF) was established in response to the 1992 Earth Summit. The IPF evolved into the inter-governmental Forum on Forests in 1997, after the UN's five-year review of the Earth Summit goals. The mission of the forum is to examine the underlying causes of deforestation and to help countries develop strategies that address them.

Efforts to advance an international legal convention on forests, which began in 1990, have been shelved, however. Some observes believe that advancing such a convention would only codify the standards of a weak consensus and thus would be worse than no convention at all. Widespread opposition to a convention makes it unlikely that the issues will reach the negotiating table.

Instead, many organizations urge governments of countries with large forest resources to enforce existing legislation and to introduce more effective forest conservation initiatives. Close to 130 countries have developed or updated their National Forest Programmes over the past decade.

While such initiatives are promising, they cannot be expected to halt forest destruction completely. Millions of people rely on forest products for their livelihoods. Sustainable forest management will require not just enforcement of laws that project forest but also alternative sources of livelihood for many rural people.

5

Ecosystems, Our Unknown Protectors

How ecosystems work and what part they play in biodiversity remain a mystery. Buy we do know that they perform a host of invaluable services for the human species. In my view, biodiversity's fundamental value is neither aesthetic nor economic but environmental, even though most people are largely unaware of this. The value of biodiversity is often measured in terms of the number of species living in a given area. But the interactions between the many species in an ecosystem, and between them and the environment's physical and chemical components are also very important. This highly intricate web of relationships makes an ecosystem more valuable than the sum of the species it contains.

Ecosystems perform services that are essential for the survival of the human species. They fix carbon in the atmosphere and produce oxygen, protect soil from erosion and keep it fertile, filter water and replenish aquifers, provide pollination and anti-parasite agents and so on.

The first two of these services are closely related to each other. They result from photosynthesis, whereby green plants, starting with algae, absorb carbon dioxide (CO_2) and emit oxygen. For millions of years, the balance between the various gases in the atmosphere remained stable. But with the coming of the industrial revolution, humans began burning increased amounts of fossil fuels. Today, three billion tonnes of carbon build up in the atmosphere each year, and natural ecosystems can no longer absorb all these emissions—especially since they are disappearing at an alarming rate. Deforestation alone releases such tremendous amounts of CO_2 and other gases, such as methane, that it has become the second-leading cause of global warming.

Storing freshwater, protecting soil and keeping it fertile are three other closely related functions. Ecosystems are veritable "freshwater factories". They absorb rainwater and slowly filter it through the soil before draining it towards streams, rivers, lakes and underground aquifers that supply us with the precious liquid. When the vegetal ground cover is degraded, the water cycle is disrupted. Rain strikes the bare earth, washing away huge amounts of nutritional substances. Reservoirs, lakes and rivers silt up.

Uncertain Reaction to Climate Change

Despite years of research, scientists still know very tittle about how ecosystems work. We are generally incapable of predicting how they will react to certain transformations in the environment, especially climate changes. Nor do we know any more about whether a species present in a given environment is superfluous or "replicable" even when it is very rate. Likewise, we do not know which key species are indispensable to maintaining an ecosystem, with a few exceptions such as pine forests, where the tree is obviously the dominant species.

We know even less about the part biological diversity itself play in maintaining ecosystems and the services they perform. One simple example is a highest diversified forest that absorbs carbon dioxide—a vital function, as we have seen, for limiting global warming. Suppose the forest is cleared to make way for a single-crop forest. The service will still be performed, perhaps even better at first because young, fast-growing trees absorb more CO_2 than old ones, which regenerate slowly. But what will happen in the long term? After several decades, the consequences of the loss of biodiversity will probably be felt. Replacing many species with a single one will have certainly depleted the soil and, in the long term, slowed down the forest's growth and consequently its ability to absorb CO_2.

More generally, diversified ecosystems seem more productive. Specialists remain wary about their conclusions, but today they believe that bio-diversity helps ecosystems to resist alien species and diseases and to recover faster in the event of disruption. In the face of doubt, and to find out more about them, it is better to preserve as many different ecosystems as possible.

A Costly Lesson for New York City

Most people take it for granted that ecosystems will carry on performing services without receiving anything in return. They think nature will continue benefiting humanity, no matter how much damage is done. The survival of organisms other than our own species is perceived as a frill that future generations can live without.

These preconceived ideas are wrong and dangerous, as the city of New York has recently come to realize. The city's water has always enjoyed such a good reputation that it was sold throughout the northeastern United States. Its equality was due to Catskill Mountains' natural purification system. But that ecosystem has suffered so much from pollution, especially fertilizer run-off from farms, that by the late 1990s New York's water had become undrinkable. The city planned to build a purification plant, whose cost was put at between six and eight billion dollars, not including the $300 million in yearly operating costs—an astronomical bill for a service that had always been free! The price was so staggering that the city eventually decided to restore the Catskill Mountains' degraded environment at a cost of only one billion dollars.

This story clearly illustrates where our interests lie. We must preserve ecosystems and the conditions that enable our planet to ensure the survival of homosapiens or, at least, the short-term maintenance of our current quality of life.

6
Climate Change

Over the last half-century, carbon emissions from fossil fuel burning expanded at nearly twice the rate of population, boosting atmospheric concentrations of carbon dioxide, the principal greenhouse gas, by 30 per cent over preindustrial levels. All major scientific bodies acknowledge the likelihood that climate change due to the build-up of greenhouse gases in the atmosphere is indeed under way. The 15 warmest years on record have all occurred since 1979 and 1998.

The destabilization of our climate threatens more intense heat waves, more severe droughts and floods, more destructive storms, and more extensive forest fires. The related shifts in rainfall and temperature may jeopardize food production, the Earth's biological diversity, and entire ecosystems, as well as human health by expanding the ranges of tropical diseases. Unless efforts to curb them are stepped up, carbon emissions will continue to grow faster than population over the next 50 years, driving the Earth's climate system into unchartered territory. The Intergovernmental Panel on Climate Change (IPCC) estimates that an eventual two-thirds reduction in global emissions is needed to avoid precariously high levels of atmostpheric carbon dioxide concentrations.

The IPCC and the U.S. Department of Energy (DOE) project that emissions from developing countries will nearly quadruple over the next half-century, while those from industrial nations will increase by 30 per cent. Although annual emissions from industrial countries are currently twice as high as from developing ones, the latter are on target to eclipse the industrial world by 2020.

Higher per capita carbon emissions account for roughly 55 per cent of the increase in emissions projected for developing nations. Emissions per person are due to more than double from 0.51 tons of carbon per year in 2000—just one-fifth of the industrial level—to 1.14 tons in 2050. The remaining 45 per cent of emissions increases is due to population growth.

Fossil fuel use accounts for roughly three-quarters of world carbon emissions. As a result, regional growth in carbon emissions tend to occur where economic activity, and related energy use, is projected to grow most rapidly. Emissions in China are projected to grow over three times faster than population in the next half-century, as emissions per person soar from 0.77 tons of carbon to 2.81 tons due to booming economy that is heavily reliant on coal and other carbon-rich energy sources. In Africa, in contrast, emissions per person are expected to scarcely change—growing from the current level of 0.33 tons in 2050, despite a threefold increase in total emissions.

The effects of population growth are most profound in countries where people are heavily emitters. For example, the 115 million people added to the population of the United States between 1950 and 1998—an increase of nearly 75 per cent in just 45 years—account for more than one-tenth of current global emissions. And the carbon emissions of the 75 million people who will be added to the U.S. population in the next 50 years roughly equal the emissions of the 1.3 billion people who will be added to Africa during the period.

Deforestation and other land use changes account for the remainder of world carbon emissions. Forests have served as a sink for carbon throughout much of human history. In recent years, however, the world's forests have become net sources of atmospheric carbon, largely due to forest burning and clearing in the tropics. Six months of fires in Asia in 1997 and 1998 released more carbon than Western Europe emits from fossil fuel burning in an entire year. The carbon contribution from this source will likely increase in coming years as the burgeoning human population continues to cut down forests.

7

Climate Change and Human Health

Changes in India's climate, stemming from the greenhouse effect, are highly lil ly to damage human health. Food and fresh water supplies will be disrupted, millions of people displaced, and disease patterns altered dangerously and unpredictably.

Human health could be affected by even quite small changes in average mean temperature, and there is the prospect of some major diseases flourishing in warmer conditions and of more resistant strains of infection emerging.

The population in india most vulnerable to the negative impacts of global warming are in the lower-income groups, residents of coastal lowlands and islands, those living in semi-arid lands, and the urban poor in the squatter settlements, slums and shanty-towns of large cities.

Present strategies for immunization, coping with disease vectors or carriers, providing safe drinking water, and improving nutrition are all based on existing climate regimes, ecosystems, and sea and solar radiation levels. These are all expected to change, but exactly how much cannot be predicted. It is, therefore, virtually impossible to adjust health and nutritional strategies to take account of possible climate changes.

Humans can adapt to moderate changes in temperature and to occasional extremes. But this adaptive capacity is relatively low in infants and the elderly; it rises through childhood and adolescence to reach a maximum which can be maintained up to about 30 years of age.

A changing climate would alter the ecosystems of the vectors or agents which carry or cause many diseases, whether these be viruses, bacteria, parasites, plants, insects or other animals such as mosquitoes and snails. As the weather warms, the boundaries of the tropics may extend into the present subtropics, and parts of temperate areas may become subtropical. As air temperatures increase, some diseases will become common in regions which once rarely knew them and where there is little natural resistance to them. As a result, death rates may also climb significantly.

It is possible that warmer weather around the world will cause increases in summer diseases and decreases in those associated with winter. Diseases contracted from both water and air will also spread more readily as ambient temperatures rise. In a warmer climate, mosquitoes and other vectors also may migrate vertically, up into highlands which were once too cold for them. This may be particularly hazardous in tropical highland areas where there is no natural resistance to malaria.

Changes in temperature, rainfall, humidity and storm patterns may affect diseases borne by vectors in two ways. First, they will directly affect the vector's reproduction rate, biting rate, and the duration and frequency of human exposure. Second, they may modify agricultural systems or plant species, thus changing the relationship between host and vector. Development rates of malarial mosquitoes, for example, increase with warmer temperatures, but these pests need wet areas in which to breed.

Sea-level rise could also spread infectious disease by flooding sewerage and sanitation systems in coastal cities, and increase the incidence of diarrhoea in children. The flooding of hazardous waste dumps and sanitation systems could lead to long-term contamination of crop lands.

Rising seas may also disrupt marine habitats land aquatic food chains. Since fish constitute 40 per cent of all animal protein consumed by the people of India such a disruption of the marine ecosystem would affect the food supplies of many millions of people and dramatically increase protein deficiency and malnutrition. Changes in the availability of food and water, as well as radical shifts in disease patterns, could initiate large migrations of people, exacerbating food shortages, overcrowding, social stress and instability.

Some of the factors contributing significantly to global warming, such as the burning of fossil fuels and the use of chlorofluorocarbons (CFCs) and halons, threaten human health in other ways too. A typical petrol-driven motor car, for example, emits carbon monoxide, sulphur and nitrogen oxides, hydrocarbons, low-level ozone and lead—all of which are hazardous to health.

The ozone-depleting CFCs and halons pose a particular threat to humans through an increased risk of skin cancer, cataracts and lower immunity to other illnesses as a result of increased exposure to ultraviolet B radiation from the sun. Skin cancer risks are expected to rise most among fair-skinned.

8

Environmental Protection

The Devil May Care

Back in the 1960s, most of us thought of the environment in terms of a medieval morality play. Thirty years later, those clear moral certainties are gone. Conservation has evolved into sustainable development (SD). And instead of denouncing the Devil of industry and all his works, the Goodies have become hopelessly ensnared in dialogue with him.

Most worrisome to all true believers, the debate between the SD movement and the private sector is proving remarkably fruitful. Indeed, this interface is the origin of many of the most interesting and provocative new ideas in sustainable development today.

The strategy for environmental protection in the 1970s relied overwhelmingly on government regulations and inspectors. Dubbed "Command-and-control" by business in involved government laying out detailed rules and using bureaucratic procedures to enforce them. Industry says that centralized hands-on regulation is inherently expensive and inefficient: the rules take so long to draw up that they impose standards that are two or more years out of date.

There are a number of reasons why command-and-control is no longer appropriate as a sole or primary strategy (except in some areas, such as nuclear power safety). Over the years, our concept of the proper role of the state has narrowed, with growing distaste for unnecessary bureaucratic control. And globalization has created entirely new possibilities for market-based environmental management. A third reason is that the growth of civil society and its empowerment through the internet have substantially increased the risks to business of egregiously bad environmental behaviour. In most countries, environmental stupidity by industry is now likely to be exposed, and public pressure is able to insist on expensive remedies.

However, one factor stands out above all: the rule-book approach to the environment is just too expensive. Even the rich industrialized countries can't afford more and more regulations, inspectors and control boards. And for the developing world, using scarce skilled workers as environmental police seems even less practical. The tendency now in the major industrialized countries is for government to set performance goals, and to let industry decide how to reach them.

You just can't *trust* guys, right? But consider how we make companies keep more or less honest books. We don't have vast numbers of accountants on the public payroll counting through the receipts of everyone from Exxon to the corner store. The state lays down the basic book-keeping rules, and makes every business pay for an independent auditor to make sure that the rules are obeyed. Classic self-regulation, at little cost to the taxpayer.

Why not do the same for pollution control, product and workplace safety? A raft of pilot schemes is now under way to do just that. For example, so far most self-regulating schemes Back transparency: they are not based on publicly available rules with published emission levels. Nor are self-regulating companies usually accountable to anyone but themselves. These defects are recognized by some in industry, and attempts to correct them are being tested. So too is the concept of having such schemes formally registered with government, with legal or political sanctions for noncompliance.

Government likes self-regulation because it is cheap. And the SD lobby, somewhat unwillingly, is coming to accept this approach, especially for the developing world and "economies in transition". If Europe and North America couldn't afford command-and-control, how can countries with far fewer lawyers, administrators, chemical engineers and lab workers? Business says that self-regulation leaves the private sector alone to do what it does best: finding the most cost-effective ways of reaching agreed goals.

Not everyone in the SD camp is happy with this increasingly close relationship. Some environmentalists question the motives of business. But in the end, these are irrelevant. What really matters is what business does.

Business motives for SD are complex, and public relations is certainly one of them. Another is staff recruitment and morale: people don't like working for a company that is perceived to be damaging the environment. There are straight bottom-line motives too: bad environmental performance can depress stock prices and cause boardroom firings. It is cheaper to design a new plant for environmental standards of tomorrow than to engage in expensive retrofits later. And reducing waste has often proved extremely profitable.

Profit. The environment and development movements used to see this as a dirty word. But as these two groups melded into a single SD community, two key realizations started to emerge. The first was that sustainability does not ultimately depend on governments but on business, because business controls more than 90 per cent of the investment that shapes our future. The second was that business is based on profit so that achieving SD will require making sustainability profitable.

Making the profit motive work for sustainability is a carrot and stick affair—but without the stick the donkey would probably not go for the carrot.

In a competitive market, a boycott that cuts sales by a mere 10 per cent can make business hungry for the carrot. One promising carrot is eco-labeling, which enables consumers to make positive SD choices. The problem is to get producers, retailers, campaigners and consumers to agree on who gets this green stamp of approval. What exactly is an energy-efficient light bulb, an organically grown pineapple of a sustainable logged armchair?

To someone reared on the view that you can't trust business at all sustainability initiatives seem to be cropping up in the oddest places. The world's insurance companies, for example, are seriously worried about global warming, not just because it involves more storms, floods and severe weather, but because these events are unpredictable. Between 1991 and 1997, extreme weather cost the insurance industry US$200 billion, and companies fear that further rapid climate change could cause widespread bankruptcies. As a result, insurance has been lobbying for tighter controls on carbon dioxide than those agreed to in Kyoto.

Business and sustainable development remain uneasy bedfellows. But what cannot be denied is that the interface between them is spawning a wealth of intriguing ideas, insights and approaches.

This does not mean that civil society, governments and international institutions should be doing whatever business wants. On the contrary, as the private sector grows beyond the power of all but the largest states, the need for governmental and intergovernmental ground rules and for grassroots and international analysis, monitoring and criticism is more urgent than ever. Market mechanisms and self-regulation are still in their infancy. But the pattern of government setting broad goals, and industry working out how to meet (or exceed) them, and being publicly accountable for doing so, will be centrally to SD in the 21st century.

9

Sustainable Tourism and the Environment

Tourism is high on the international agenda. The 7th session of the Commission on Sustainable Development focused on tourism and subsequently work programmes on sustainable tourism are being developed. Also the Convention on Biological Diversity is embarking on tourism programmes and bilateral and multilateral financial institutions placed tourism high on their priority lists. The UN declared 2002 as the International Year of Ecotourism and the World Tourism Organization adopted a Global Code of Ethics for Tourism at its General Assembly, held in Santiago de Chile.

The World Tourism Organization forecasts that there will be 702 million international arrivals in the year 2002, the arrivals will top 1 billion in the year 2010 and that by 2020 international arrivals will reach 1.6 billion—nearly three times the number of international trips made in 1996, which was 592 million.

Travellers of the 21st century will go farther and farther. The Tourism 2020 Vision forecast predicts that by 2020 one out of every three trips will be a long-haul journey to another region of the world. It is expected that China will become a major force in international tourism and the WTO predicts that about 100 million Chinese will take international trips by 2020, thus putting them in fourth place in numbers of travellers after Germany, Japan and the United States. By the same time, China will attract 137 million visitors—63.5 million overseas visitors travelled to China in 1998 and thus outrank France as the world's top destination. It is estimated that during 1999 France will receive a record number of tourists of more than 70 million; in 2007 France hopes to attract 90 million visitors. The key resource for the most popular tourist destinations is the most popular tourist destinations is the natural environment: coastal resorts, tropical rainforests, wildlife in national parks and alpine skiresorts, all rely on a mixture of natural beauty, good weather and safe conditions to attract holiday destination is landscape and natural environment, followed by climate, the cost of the journey and the historical features of the place to visit. Hence, conserving the ecological integrity and environment is imperative if tourism is to be sustained.

The pressure from millions of tourists on water and marine resources, on land and landscape, on wildlife and habitat is enormous and often has devastating impact on the environment and the local population who are increasingly deprived of access to clean water and other natural resources.

In some regions, particularly in small island countries, tourism is one of the major reasons for wasting and polluting water: on average one tourist consumes at least 6 times more water than a local resident

Major water wasters and polluters are golf courses. In many countries, golf has brought heavy ecological and social costs: deforestation, the destruction of biodiversity and erosion; dispossession of peoples' homes and farms; over-consumption and pollution of water and very high use of pesticides and fertilisers which threaten local residents, workers, wildlife and the golfers themselves. A survey by the Japanese National Doctors Health Insurance Association has revealed that many golfers, caddies and residents living near a golf course suffer from skin inflammation, disorders of the ear, nose and throat and other respiratory illnesses to the inhalation of pesticides because up to 90 per cent of the chemicals sprayed on golf courses end up in the air. In some areas in Thailand, diseases emerged which, prior to the construction of golf courses, had not been known.

In some regions, golf courses have depleted water supply, agricultural production has come to a half, peasants have become impoverished and forced to migrate to urban areas in search of employment. Golf courses take large amounts of land. It is estimated that each year worldwide up to 5,000 hectares of forest are cut to clear land for golf courses.

Very often, the construction of golf courses forms an integral part of a comprehensive tourism project. Adjacent to the golf course, condominiums and/or hotels are built, very often also a marina, an airport and a casino. Studies have shown that such a complex not only has touristic objectives but is often connected to drug trafficking and money-laundering. Even the US State Department has emphasised the like between tourism, money-laundering and offshore banking.

Cruise ships are a major cause for pollution in the Caribbean, destroying maritime life and reefs by releasing waste into the ocean. Recently the Royal Caribbean, the world's second largest cruise line, was fined a record sum of US$ 18 million for dumping waste oil and hazardous chemicals into the sea. The company admitted to routinely dumping wasted oil from its fleet and that it deliberately dumped in U.S. harbours and coastal areas many other types of pollutants, including hazardous chemicals from photo processing equipment, dry cleaning shops and printing presses. Some hazardous materials, including toxic solvents from dry cleaning operations, were illegally places in the garbage aboard the ships. The material was then either incinerated on the ship or dumped in U.S. or foreign ports mixed with ordinary garbage.

It was announced that the Royal Caribbean Cruise reported a profit of US$ 338 million in 1997, a 93 per cent increase over the previous year, Carnival Corporation's Holland, the biggest cruise company with a turnover of US$ 3 billion in 1997 made a net profit of US $836 million, 25 per cent more than in 1996. Both cruise companies have recently been fined millions of dollars for dumping untreated bilge water, oil and other waste into Alaskan waters.

However, the impact of oil and hazardous waste on water, maritime life and coral reefs is devastating and all fines paid for the damage caused by the cruise ships will not revive dead corals.

A recent Green peace study on coral reefs—one of the marine world's great natural treasures—predicts that the coral bleaching which dramatically whitened many of the world's reefs last year will escalate rapidly under accepted global climate models and that the damage would wreak havoc in fisheries and tourism, disrupting the economies of many nations.

A WWF study recently published on "Climate Change and its Impacts on Tourism," warned that droughts, rising seas, flash floods, forest fires and diseases could turn profitable destinations into holiday horror stories. The report urges the tourist industry to persuade western industrialised governments to take more concerted action to reduce their nations' carbon dioxide emissions the main cause of global warming.

The Need for Action and Education

If governments, the international community and the tourism industry want to save the world's major tourist destinations, immediate action is required. Governments and the tourism industry must abide to the principle that environmental protection is an integral part of tourism development. In order to protect the environment and mitigate the damages caused by tourism, some countries have decided to take action: The Spanish island Minorca and the Seychelles will introduce Eco-tax on tourism. This tax will be around US$ 12 per person in Minorca and its revenues are earmarked for the maintenance of national parks and the restoration of damaged coastline. Visitors to the Seychelles will have to buy a so-called "gold-card" at a price of 100$ which entitles unlimited access to the country; income from this card will be used for sewage management and protection of fresh water supply.

Only if tourism investors and developers:

(a) consider the natural capacity for the regeneration and future productivity of natural resources:

(b) recognise the contribution that people and communities, customs and life styles, make to the tourism experience and therefore accept that these people must have an equitable share in the economic benefits of tourism: and

(c) listen to local people in the tourist destinations, tourism may become sustainable.

Education and awareness raising campaigns at all levels are therefore imperative.

10

Sustainable Tourism Development

Tourism has gown into one of the world's major industries and has thus also become an increasingly important, if complex, issue for environmental policy. Unless it is developed in a sustainable manner, we will be unable to achieve key objectives of global environmental policy such as the preservation of biological diversity, the prevention of climate change or the conservation of natural resources.

Tourism itself depends a lot on the existence of unspoilt nature and landscapes, as well as a healthy environment. If nature is plundered, landscapes are destroyed or water, energy and soil resources are over-exploited, the economic basis of tourism is also undermined. The needs of tourism do therefore overlap with those of environmental protection and nature conservation.

On the one hand, for instance, tourists are becoming increasingly environmentally conscious and are looking to get back to nature and enjoy unspoilt environments when on holiday. On the other hand, however, the number of international tourists is growing constantly. The proportion of long-haul journeys is also increasing steadily, especially in the industrialised nations, where travel is now taken for granted as part of people's liftestyles and has become an important factor in social status. The many different types of travel and holidays are covering more and more countries and regions and, as a result, increasing numbers of previously unspoilt natural environments are being opened up to tourism. This applies equally to coastlines, small islands, coral reefs, rock formations and mountain regions.

There is growing recognition of the need for tourism to develop in a sustainable and environmentally friendly manner. Many countries have, for instance, introduced regulations which require environmental impact surveys to be carried out at least for larger tourist developments. Since the Rio Summit in 1992, there have also been more initiatives in support of sustainable tourism at international level.

- Sustainable tourism allows for the rational use of biological diversity and can contribute to the preservation of that diversity.
- The development of tourism must be controlled and carefully managed so that it remains sustainable.
- Particular attention must be paid to tourism in ecologically and culturally sensitive areas, where mass tourism should be avoided.
- All parties concerned, including in particular the private sector, have a part to play in bringing about the sustainable development of tourism, and voluntary initiatives (codes of conduct, quality labels) should be encouraged.
- Particular importance should be attached to the local level, which is not only responsible for the sustainable development of tourism but should also derive particular benefit from tourism.

It will mark the successful beginning of internationally coordinated efforts to make tourism environmentally and socially sustainable so that many generations to come can continue to experience and enjoy the beauty of nature on our planet.

11

Sustainable Cities

Today almost one half of the world's population lives in cities. The world's cities are growing by one million people each week. Cities today play a significant role in development. They continue to attract migrants from rural areas because they enable people to advance socially and economically. Cities offer significant economies of scale in the provision of jobs, housing and services, and are important centers of productivity and social development.

However, the stress of this rapid urban population growth is often overwhelming. The long list of afflictions includes urban poverty rates of up to 60 per cent. Despite growing investments, more than one-third of the urban population live in substandard housing. Forty per cent of urban dwellers do not have access to safe drinking water or adequate sanitation. Primarily due to a rapid growth and a deteriorating urban environment, at least 600 million people in human settlements (cities, towns and villages) already live in health- and life-threatening situations, and almost 50 per cent of these are children.

The high rate of urban population growth in most regions has led to common problems: congestion, lack of funds to provide basic services, a shortage of adequate housing and declining infrastructure, to name a few.

While these problems are occurring in urban areas, cities still have an important role to play in protecting the global environment in the face of rapid urban population growth. Agricultural and livestock production in rural areas is pushing farther and farther into ecologically fragile regions and cannot support growing population. The finite land and water resources make it imperative that human settlements be carefully planned. Indeed, sustainable urbanization will ease the pressures caused by encroachment of fragile natural habitats.

India's cities offer a bewildering sight to any visitor: the congestion caused by rapid population growth and a continuing rural-urban draft often leads to conditions which defy all rules of orders, hygiene and environmental safety. Inadequate leadership, corruption and mismanagement have a harmful effect on the physical, environmental, social and ethical structures of cities in India.

Millions of people live in inadequate conditions—without piped water, electricity, security of land tenure, access to roads or health facilities. The means available for production and financing of housing and urban infrastructure are too limited to meet basic needs.

Reducing Poverty and Creating Jobs

Urban poverty is rising at an alarming pace, especially among women. The informal economic sector—which makes a substantial contribution to the delivery of services, production of goods, building of infrastructure and housing construction—often provides the only opportunity for the urban poor to make a living.

Local informal housing construction, for example, generates up to 20 per cent more jobs than high-cost construction. Street hawking, waste recycling and food production and primary sources of income among the urban poor are illustrative of the creativity of survival strategies.

However, the informal sector itself is often highly exploitative and fails to raise people's economic development beyond mere subsistence. Larger economic strategies and more participatory urban planning approaches that take stock of local skills, technologies and materials are required to generate new and better-paying job opportunities in cities and towns.

Incorporating Environmental Concerns

In 1992 the Rio Conference on Environment and Development designed the Agenda 21 Programme of Action to help save a planet endangered by environmental neglect and plagued by poverty and underdevelopment. Most of the goals agreed to in Rio can become reality only through local action in cities where environmental threats are increasing. Again, it is the urban poor who are particularly endangered by environmental degradation and pollution. The world's Agenda 21 will fail if the city's environmental agenda (pollution, inadequate sanitation, water supply and waste management) is not addressed. This is being recognized by local authorities all over the world.

Sustainable development in the twenty first-century will to a large degree, depend upon how cities, towns and villages everywhere interact with the environment and utilize natural resources.

Increasing Awareness of Gender Issues

Women and men use and experience cities differently, according to their roles, responsibilities and access to resources. For example, when basic services are lacking in a settlement, more often than not it is women who take on responsibilities such as water collection and refuse disposal. Women often have unequal access to resources such as property, credit, training and technology. All of these factors must be addressed urgently, as they make it harder for women to improve their living standards and those of their children.

Disaster Mitigation Relief and Reconstruction

As cities become large and more densely populated, they become increasingly vulnerable to natural and man-made disasters such as earthquakes, floods, industrial hazards, epidemics, civil strife and wars. Poor people are forced to live in the most exposed, dangerous and cramped conditions; in flood-prone areas, on steep hillsides or near polluted streams and waste dumps. As a result, they are most likely to lose their homes or their lives when disasters occur. Better planning, access to affordable urban land, and improved construction methods can reduce the extent of catastrophes.

These successful and sustainable approaches to poverty eradication; managing the urban environment; providing access to land, shelter and finance; empowering women and men; and many other issues will have to be decumented and disseminated widely.

12

Towards Healthy Cities

More than a third of the urban population in developing world live in housing of such poor quality with such inadequate provision fo: water, sanitation, drainage, garbage collection and health care that their health is constantly under threat. But, properly planned, cities can be safe and healthy.

In the cities of India, it is common for one child in three to die before the age of five and for virtually all infants, children and adults who survive to have disease burdens many times higher than they should.

Diarrhoea, tuberculosis and respiratory infections (each among the largest causes of death) are generally much increased by over-crowding. Many accidental injuries happen when there are three or more persons living in each small room in shelters made of flammable materials and there is little chance of providing occupants (especially children) with protection from open fires or stoves.

But cities also include some of India's safest and most healthy neighbourhoods. High densities allow much lower costs for supplying each household with piped, treated water supplies and most forms of health, educational and emergency services.

Sanitation and drainage may be costly in cities, as complex systems are needed to cope with high densities and large population concentrations but city households can generally afford to pay more—and are prepared to do so if they get a good service.

Cities may be considered ecologically unsustainable because of high consumption and waste levels but well planned and managed cities can combine high living standards with remarkably low levels of energy consumption, resource use and wastes. The concentration of people and production creates many more possibilities of collecting and recycling wastes and for walking, bicycling and a high quality public transport.

For many, city-life is one of excessive workloads and drudgery, yet cities remain centres of culture—including the visual and decorative arts, music, dance, theatre and literature. Most cities have a large reserve of young people on whose initiative and energy they could draw to improve conditions—yet most such people find that their cities offer them little hope and little prospect of employment. If cities have such potential to provide healthy, stimulating and valued places to live the work for all age groups, why do so few achieve this?

Supporting Change

Much of the explanation is the lack of 'good governance' Good governance in any city means encouragement and support from all levels of government for a great range of investments of capital, expertise and time by individuals, households, communities, voluntary organizations and NGOs—as well as private enterprises. In most cities in India, the total value of investment made by people in their own homes and neighbourhoods exceeds many times the total value of capital investments made by city and municipal authorities. Yet governments and aid agencies usually ignore (or deem illegal) most such efforts.

Most households who want their own home cannot afford to purchase one—or at least one that is legal. They cannot obtain housing loans so the cost of the house purchase can be spread over a number of years—as they cannot meet the (usually) inappropriate conditions set by banks or housing finance institutions. If they turn to building their own home—as most do—they have to occupy or purchase the site illegally. They often have to build on dangerous sites—in floodplains or out slopes with frequent landslides or mudslides—as the cost of safer sites is too high.

Even if they can qualify, for a housing loan, most such loans are for finished houses, not for incremental construction. And even when they have developed their own home and neighbourhood into a viable residential area, governments usually refuse to provide these with roads, water supplies, drains and other essential infrastructure, because they are 'illegal'.

What would cities look like today if governments had supported these individual and community efforts by ensuring that land, building materials, credit and technical advice were as cheap and readily available as possible? Or if government—community partnerships had been formed to, at least, improve water supply, sanitation, drainage and health care.

These work within what is often called the 'social economy'—the great variety of initiatives and actions that are organized and controlled locally and that are not profit-oriented. The social economy includes the work of citizen groups, residents' associations, street or barrio clubs, youth clubs, and parent associations that support local schools. It includes many voluntary groups that provide services for the elderly, the physically disabled or other individuals in need of social. It often includes many initiatives that make cities safer and more fun—helping provide supervised play space, sport and recreational opportunities for children and youth. It may provide formal or informal supervision or maintenance of parks, squares and other public spaces.

The social economy not only 'gets things done' but also creates a dense fabric of relationships that allows citizens to work together in identifying and acting on local problems. Its value to a 'healthy city' is enormous, even if it is often forgotten by governments and international agencies.

The capacity of city authorities to govern is not the same as the capacity to invest, since these authorities can do much to encourage and support the social economy. City authorities can often greatly increase the supply and reduce the cost of land for housing by changing inappropriate regulations, streamlining planning and land use control procedures and making better use of publicly owned land.

City authorities should also have the main role in enforcing legislation on air and water pollution and occupational health and safety. This does not require large investments by public authorities, but it can do much to improve health and the quality of life in a city. Good governance also means managing competing claims and finding common ground between enterprises, trade unions and residents about what should be done to make the city more healthy.

Achieving a healthy city needs a representative political system through with the priorities of citizens and businesses can influence policies and actions. Democratic structures remain among the best checks on the misallocation of resources by city and municipal governments. Actively involving a wide range of local groups in developing 'city governance' helps ensure that the different priorities of a wide range of groups are addressed.

The key issue is not so much identifying what should be done to achieve more healthy cities. This is well known. It is identifying how it should be done, especially how governments and international agencies can support a vast range of activities by individuals, households and communities that help build and maintain healthy cities—which to date they ignored or even (for many governments) repressed.

13

Tourism and the Environment

The relationship between tourism and the environment is obvious, and is largely established through what is sometimes called "environment quality". This quality is perceived in different ways according to the human population and the circumstances presiding tourist activities at any given moment. Any analysis of the relationship between tourism and the environment that we can include under human ecology therefore comprises aspects of the natural sciences as well as the social sciences.

Tourist activity is promoted, conditioned and influenced by the environmental circumstances of each region and can be affected by modifications or changes in those circumstances. Although a lot of emphasis has been placed on the negative impact or modifications in "environment quality" attributed to tourism, it is also accepted that it can be a very important factor in the preservation and defence of ecological values threatened by more destructive alternatives for the use of territory. Very often, tourism can be the most suitable and most satisfactory way of using a region's renewable natural resources. Nevertheless, their management and use need to be properly regulated so as to guarantee their renewability and persistence.

There is room in this complex field of relations to study, rationalize and optimize an activity as important as tourism, from the point of view of its insertion in the ecological systems with which it interacts. However, there are relatively few efficient studies on issues of real importance. It is startling to observe that places with tourist potential undertake little or no research in this field.

One possible cause is the difficulty in identifying the real problematic in tourism/environment relations, which is essentially interdisciplinary and involves the integration of traditionally separate areas of knowledge. Although work is undertaken from time to time on environmental psychology, the sociology of tourism, behaviour in relation to the environment, etc., they are very rarely combined with works on the environment, forestry and agricultural policies, soil use, contamination, biodiversity, evaluation of environmental impact, nature conservation, etc., in search for a more integrated management of tourist resources.

Responsible Tourism

Tourism runs the risk of going the way of other phenomena, which first of all experience rapid growth and then suffer a spectacular collapse, what in economics is often called "boom and bust".

The causes are familiar: a certain dose of greed, often based on a lack of mid- or long-term planning, property speculation, little consideration for local populations—in both economic and social aspects—and, in general, a lack of awareness as regards environmental aspects—contamination, water use, energy, etc.,— on the part of tour operators, hoteliers and other agents involved in tourism in its different forms, including the tourists themselves. The problem is particularly evident in ecotourism, based on the wonders of the natural world: landscapes, flora and fauna. Many experts fear for the future of this type of tourism, which has grown spectacularly in the last few years. Landscapes deteriorate, the fauna decreases, the designers and administrators of tourist developments fail to respect the most elementary principles for adapting architecture to its surroundings, or also there is little effort to recycle, economize or educate with a few honourable exceptions, tourist planning is careless and irresponsible.

And yet a responsible approach would be in the tour operators' own interests, as it would make the tourist industry sustainable, with positive influences on biological, economic and social aspects.

Ecotourism, for example, has shown that when properly conceived it can become a powerful instrument for the preservation of nature, with very favourable repercussions for local populations and for educational programmes, while offering hundreds of millions of ecotourists a wide range of spiritual and physical satisfactions. At the same time, the host countries can take pride in what they have to offer their citizens and the rest of the world.

The preventive and corrective measures are known to us; what is needed is a sense of responsibility and farsightedness on the part both of the authorities and of the industry. We need regulations and controls, so as to put the people who do the damage out of circulation and reward those at the forefront of sustainability.

Sustainable Tourism

After several decades of rapid quantitative growth, tourism is going through a period of profound transformation. Tourists, the consumers in this industry, but also the public, have started to demand a change in the conditions of production and use of tourist services, putting an end to the uncontrolled expansion of mass tourism.

This is the ultimate reason, apart from ethical and aesthetic considerations, why tourist activity as a whole, in the private sector as well as in the public and voluntary (NGO) sector, has begun to seriously analyse the implications of tourism in terms of socio-cultural and environmental impacts, and to consider the need to draw up and implement environment-friendly tourist policies.

Indeed, while not denying the viability and the utility of alternative approaches of an external and coercive nature, it is obvious that the decision-makes in the sector react better to positive stimuli. The realisation that their clients prefer well-conserved areas and non-aggressive tourist practices and that they are prepared to pay more for this makes it easier to adopt strategies of sustainability in the tourist industry in a sincere alliance with conservation movements.

All this points to the validity of Overall Quality Management as a viable method in sustainable tourist activities.

The overall quality approach renders the management of products and especially of tourist areas extremely sensitive to the preferences and expectations of consumers. The private public profitability of a tourist destination will depend on clients' satisfaction, since these will return more often and for longer and will pass on a positive image of their holiday experiences. Insofar as these preferences and expectations include the demand for unspoilt settings, consumer satisfaction, and therefore the profitability of a tourist spot, will call for the development of strategies for sustainable development.

One can believe this is a productive approach for sustainability in the tourist business and one that makes for professional attitudes that fit in with the economic targets of businesses and other organisations. There is only one prior requirement: continued education and training of everyone involved in tourism, from consumers to those responsible for tourist policies. The demand for quality, and even more so for environmental quality, is a call to people's awareness, to their understanding of the environmental and cultural implications of any activity and their ability to express themselves and to organise to choose the most clear-sighted line of action.

Tourism in the Modern Age

What will the tourist trade of the year 2000 be like? Who will be the tourists of the coming millennium? These are the questions which, faced with the extraordinary boom in tourism, experts, tour operators and politicians have repeatedly posed over the last fifteen years. These questions arise either because of the financial profits the tourist industry involves, or from the demands of consumers who show new awarenesses, habits and lifestyles. In the eighties, mass tourism gradually changed and people began to talk of "tourisms". Expressions such as cultural tourism, sports tourism, religious tourism, adventure tourism or ecotourism have become part of everyday language. In the past, the dominant practice was to take one long holiday in a single destination; today, people tend to distribute their holidays over different destinations and different times of the year.

From a socio-historical point of view, three types of tourist industry can be differentiated. In the case of the industrial tourist, for whom work is the center of existence, the motivations for travelling can be summed up as rest and freedom from responsibilities. This type is gradually decreasing in number. The hedonistic tourist belongs to the generation that discovered entertainment and consumerism. They like to go on holiday to experiment, to explore the unknown, enjoy themselves meet, other people and relax in unspoilt natural surroundings. These are the majority today and will continue to be so. Finally, the modern age tourist, someone who tends to reduce the polarity between work and play: not just work, but just not fun, either. Their reasons for travelling include broadening their personal horizons and getting back to simple things and nature, with a touch of creativity in the planning of their journey. These are gradually growing in number and in future will form an important segment of demand.

One characteristic in the expectations of the modern age tourist is the capacity to make a critical appraisal of the offer and to influence it. Producers should be more attentive and sensitive to the new demands and be flexible enough to cater for the tourist in search of higher quality. In the third millennium in fact, the concept of quality will have to take environmental aspects more into account. Recent forms of tourism point to a renewed interest in nature and a wish for quality tourism. So much so that some tourist spots are reorganising their own offer in keeping with these trends. Quality is the result of a complex strategy which is organised day by day. The consumers, whose environmental awareness is constantly growing, will expect to identify, verify and be able to differentiate ecologically correct products from the imitations now invading the market.

The present millennium is coming to an end and is leaving Western countries with a high level of welfare and a large tourist demand to satisfy. Nevertheless, serious environmental problems also plague areas that receive a high influx of tourists. Tourists, tour operators, local authorities and the general public are therefore called on to find new forms of coexistence and the right solutions for themselves and for the survival of the planet.

14

Urbanization and the Environment

Is abandoning the cities the answer to the growing ecological problems of urbanization? The trend at any rate is in the opposite direction. At the beginning of this century, only every 10th person worldwide was a city dweller. At its end, more than half the global population will be urbanites. And most of the urban population growth will take place in the developing countries, led by Asia.

Compared to other parts of the world, however, the urbanization process in Asia is currently not even particularly far out in front. Worldwide, city dwellers account for 43 per cent of the total population. Industrial nations have an average urbanization rate of 72 per cent. Less industrialized countries have 34 per cent. In the Asia-Pacific region the rate is 30 per cent, in Latin America 72 per cent, and in Africa 33 per cent. The urbanization growth rate in a number of Asian countries has in fact slowed compared with earlier years. Nevertheless, not only industrialization, but also the increasing degree of urbanization has emerged as a growing burden on the environment in many Asian countries.

Changed Urbanization Pattern in India

Environment burdens are just as much a problem in the old industrial nations as they are in India. But each group has a specific pattern of development. The urbanization process in India has proven to be more pollution-intensive than that in the old industrial nations of Europe and North America. There are seasons for that:

- industrialization in India is restricted to a few locations which are often concentrated in and around capital cities. Although environmental damage continues to be minor at a national level, these locations have higher pollution levels than those ever reached in industrial nations;
- furthermore, besides the strong regionalization of industries, the industrialization pattern of india shows a great diversity of environmental hazards. The trend to establish "last industries first", which is promoted by progressive industrialization, leads to a country producing certain dangerous materials before they have been covered by state regulations;
- the time factor has to be seen as a important element in the emergence of these already highly regionalized environmental burdens. In India industrialization and its concomitant urbanization is taking place within a ban population grew tremendously.

Growing Environmental Damage

Water pollution in India is caused mainly by domestic sewage. For example, households are responsible for 75 per cent of the pollution of the rivers. The domestic sewage problem got more and more out of control with growing urban populations. Pipe-based waste water systems are rare in this country. In India dealing with waste has an extremely low priority. The type of waste disposal depends mostly on what the cities can afford. The present level of air pollution is also very high.

Innovative Approaches to Solutions

Environmental protection and economic development are seen as contradictions. Economic development can only be achieved at the cost of higher levels of environmental pollution. And in reverse, if pollution is to be controlled and reduced this can only be done to the disadvantage of further development. In the meantime, however, numerous instances of successful urban environmental management are developing. They could help to change and subsequently break through the existing pattern of thinking. The following approaches can be viewed as important.

Combining regulations with incentives: The introduction of lead-free petrol and the mandatory equipping of new cars with catalytic converters is still by no means common in India. As numerous cars without catalytic converters are still able to use lead-free. Converters were then at first made compulsory for higher-powered cars, and later also for compact models.

Combining regulations with simple controls: Apart from general limitation of the number of cars in the city, its most important single measure to prevent traffic jams and the additional petrol consumption and pollutant emissions caused by them.

High economic growth in India has in fact led to a general reduction of poverty. But the distribution of income, particularly between urban and rural areas, has remained relatively constant. Urban environmental and traffic problems have increased heavily during the same period. These developments can be attributed to a certain pattern of official action (or "non-action"):

- governments have made efforts in supplying roads, but neglected the demand for mobility.
- governments are preoccupied with supplying water, and have neglected follow-up problems, above all the questions of waste water disposal and treatment. In Indian cities, for example, this leads to the absurd situation that due to the mushroom-like growth of the cities and the increased water pollution linked with it, water must be brought in over ever greater distances and at ever greater expense;
- governments take a one-sided look at noxious substances. Concentrations of harmful substances in water and in the air are in fact checked, and some measures are taken against individual pollutants of single sectors (e.g. lead emissions by the transport sector).

But an integrated policy which operaters integrated environmental management with the aim of comprehensively relieving the burdens on the environment has not yet been developed anywhere. To consider such a concept, it is necessary to cut loose from the customary way of approaching problems. It makes sense not to separate the problem areas from each other according to sectors and pollutants, but rather on the basis of their ecological impact.

Orienting on demand hits the core of the concept of ecological modernization, which is about reducing the intensity of resource use (note, at this stage this does not yet mean the absolute reduction of inputs). At the same time, sights are set on a lower use of land with the same size of population, or also lower energy consumption with the same degree of added value or the same per capita income.

Finally, the importance of governments for creating framework conditions must be emphasized one again. Because the actors come from different spheres, such conditions are essential.

From the time of the Greek polis, it was the ambition of the Greek city councillors to pass on a city that was more beautiful than the one they had taken over. There is a long way to go before such an attribute asserts itself in India (and elsewhere).

15

The Environment, the Economy and Public Health

AN INTEGRATED VIEW

The environment is central to the health of people and their economies. Just as a foetus is totally dependent on the life-support system of the mother during her pregnancy, so the health and vitality of people and their economies are totally dependent on their environments. Unfortunately, many people do not see it that way. The either see the environment as dependent on the economy—such as the politician who says: "let's make the economy strong, then we'll fix the environment when we can afford it"—or they see little connection between health and the environment, whether they are "deep greens" campaigning on ecological issues or doctors treating individual patients and individual illnesses. Whether we are politicians, greens or doctors, is there not a more efficient way to fulfill our aims? For this, a broader perspective is essential.

All economies are sub-systems of the larger environmental system which provides the:

- Sources of energy and materials;
- Sinks for pollution and other wastes;
- Services of water, nutrients and carbon recycling.
- Space for living, working and aesthetics ("a walk in the woods and the song of a bird")

Neglect of this life-support system of the "4 S's" leads to weaker or defunct economies as vegetation, food soils, water or air become contaminated or exhausted and gradually fail to support economic activity. This is dramatically illustrated in the Aral Sea region, or the collapsed Canadian salmon fishing communities.

Indirect Social Costs

Less catastrophic but still costly is where economic damage is caused by pesticides and nutrient contamination of groundwater, involving millions of Rupees in water treatment. This is a social cost to the economy that the agricultural sector does not include in the price of its food: an economic distortion that reduces the real wealth of society via false price signals that encourage the over-use of pesticides and fertilisers. Similarly, the "external" costs on society of road-respiratory-induced accidents, noise, respiratory and circulatory diseases and congestion amount to a lot of money to any government but these costs are not borne by transport users, which mean that transport is encouraged beyond the level that is economic for society as a whole. By internalising these externalities via taxes and other means, the market prices for transport would become fairer and more efficient. Currently only about 30 per cent of transport externalities are covered by transport taxes. But if the health of an economy is dependent on the health of its environment, what about the health of its people?

Without access to the basics of clean water, shelter, fresh air and food, people obviously suffer. Even in more developed economies where the link between everyday life and the environment is not so visible, the role of environmental factors in disease and well-being is significant. Most of the major diseases such as heart disease, concer, respiratory diseases and allergies have an environmental as well as a genetic component within a multi-factorial chain of causation. And while each environmental factor may be small, if the links in the chain of causation are inter-dependent, as they often appear to be then removing even a small link can break the chain.

Environmental Factors

Take asthma in children, for example. There seem to be many causes, from a child's genetic inheritance to its nutritional status, which in turn help determine how it reacts to the many environmental factors, both indoor (such as mites, pets, damp, environmental tobacco smoke, nitrogen oxides) and outdoor (such as pollen and pollution from industry and traffic), that have been implicated in asthma causation. Therefore it is clear that diagnoses of asthma and many other diseases should systematically embrace environmental factors. This will be a significant challenge for doctors whose time is scarce and whose training is not usually appropriate.

This multi-causal chain will very in its exact make-up from child to child, but for children overall, even if the environmental factors such as damp housing to traffic fumes may be less important than, say, genetic make-up or nutritional status, the environmental factors may be the ones that can be most cost effectively removed, thus breaking the causal chain. And, as with many environmental issues, there are secondary benefits of action, such as less noise or fewer accidents from traffic reduction, or energy savings from dry houses, which further justify the environmental actions even where exact causations are not well understood.

The environmental causes of disease and ill-health are a controversial and ill understood area of science and opinions vary about their significance. Some say that, for Western Europe, perhaps 2-3 per cent of public disease and ill-health is determined by known environmental factors but others maintain that it must be far more significant. They point to the sharp increase over the last two or three decades in asthma, allergies, and cancers (particularly of the reproductive organs such as breast and testicles) and related ill-health such as sperm count decline, which cannot be explained by genetic causes. They also observe that the large differences in health between the socio-economic classes cannot be explained without involving significant environmental causation.

It is thought that the ubiquitous presence of low doses of mixtures of chemicals in food, drink, air, consumer products and the general environment are playing some role in public ill-health, even if the evidence for this is far from substantial.

Impact on Public Health

But what bout environmental programmes and campaigns being little concerned with health? Well, history so far shows that the environment only gets serious attention when it is seen to be damaging either the economy or public health. Yet because "everything connects" in "socio-enviro" systems, action to stop infectious diseases from water contamination, or to reduce skin cancer from ozone depletion, leads to a better environment for all species. And if upland forests are preserved because they are seen to be cheaper and more effective water regulators (which reduce the risk of lowland flooding) than dams, them upland biodiversity benefits anyway, even if it was last in the queue for political attention.

Although public health may be seen by some as only a small part of "the environment", much environmental progress depends upon the political weight of the health impacts. For example, the cost benefit exercise on the current multi-pollutant/ effect programme on acidification, eutrophication and low-level ozone shows that it is the benefits to human health, not eco-system damage, that provide the main economic justification for further reductions in SO_2, No_x and Nh_3. Ecologists need the language of public health in order to maximise political support for the environment. So, it is out of our specialist "boxes" of economics, health and ecology, and into a shared systems approach, with integrated programmes that build partnerships for progress.

16

An Agenda for Change

The world's growing population, combined with unsustainable production and consumption patterns, is putting increasing stress on air, land, energy, and other essential resources.

- Development strategies will have to deal with the combination of population growth ecosystem health, technology, and access to resources. Meeting the unmet need for family planning and reproductive health services should be part of national sustainable development strategies.
- The world needs to do a better job of forecasting the possible outcome of current human activities, including population trends, per capita resource use, and wealth distribution.

Protecting the atmosphere. The atmosphere is under increasing pressure from green house gases that threaten to change the climate and from chemicals that reduce the ozone layer. Governments need to:

- modernize existing power system to gain energy efficiency and develop new and renewable energy sources.
- promote national energy efficiency and emission standards and develop efficient, cost-effective, and less polluting mass transit systems.

Combating deforestation. Forests worldwide are threatened by uncontrolled degradation and conversion to other uses because of increasing human pressure.

- There is an urgent need to conserve and plant forests in developed and developing countries to maintain or restore the ecological balance and to provide for human needs.
- Governments need to work with business, scientists, local community groups, indigenous people, and the public to create long-term conservation and management policies for every forest region and watershed.

Sustainable agriculture and rural development. Hunger is already a constant threat to over 800 million people, while the world's ability to continue meeting growing demand for food and other agricultural products over the long term is uncertain. Soil erosion, salinization, waterlogging, and loss of soil fertility are increasing in all countries.

Agriculture has to meet rising needs mainly by increasing productivity, because most of the world's best croplands are already in use. At the same time further encroachment on land that is only marginally suitable for cultivation must be avoided.

- Sustainable agriculture and rural development will require major adjustments in agricultural, environmental, and economic policies in all countries and at the international level.

Conservation of biological diversity. The loss of the world's biological diversity continues, mainly from habitat destruction, over-harvesting, pollution, of foreign plants and animals (known and exotics). This decline in biodiversity is largely caused by human activity and represents a serious threat to our development.

- Develop national strategies to conserve and sustainably use biological diversity and to make these strategies part of overall national development efforts.
- Implement fair sharing of the benefits between providers and consumers of biological resources.
- Protect natural habitats. Promote the rehabilitation of damaged ecosystems.

Protecting and managing the oceans. Oceans are under increasing environmental stress from pollution over-fishing, and degradation of coastlines and coral reefs. Abut 70 per cent of marine pollution comes from sources on land. Countries should commit themselves to control and reduce degradation of the marine environment. They should:

- build and maintain sewage-treatment systems and avoid discharging sewage near shell fisheries, water intakes and bathing areas.
- develop land-use practices that reduce run-off of soil and wastes to rivers and thus to the seas. Use environmentally less harmful pesticides and fertilizers.
- control and prevent coastal erosion and silting due to land uses such as unplanned constriction.

Protecting and managing fresh water. In many parts of the world there is widespread scarcity, gradual destruction, and increased pollution of freshwater resources. The causes include the inadequately treated sewage and industrial waste, loss of natural water in catchment areas, deforestation and other chemicals into the water. The following approaches are key:

- The way to provide all people with potable water and basic sanitation is to adopt the approach "some for all rather than more for some." This approach can be achieved through low-cost services built and maintained at the community level.
- Notions need to identify and protect water resources and see that water is used on a sustainable basis. They need effective water pollution prevention and control programmes. There is a particular need for appropriate sanitation and waste-disposal technologies for low-income, high-density cities.

17

Population and the Environment— The Global Challenge

As the century begins, natural resources are under increasing pressure, threatening public health and development. Water shortages, soil exhaustion, loss of forests, air and water pollution, and degradation of coastlines afflict many areas. As the world's population grown, improving living standards without destroying the environment is a global challenge.

Most developed economies currently consume resources much faster than they can regenerate. Most developing countries with rapid population growth face the urgent need to improve living standards. As we humans exploit nature to meet present needs, are we destroying resources needed for the future?

Environment Getting Worse

In the past decade in every environmental sector, conditions have either failed to improve, or they are worsening:

Public health. Unclean water, along with poor sanitation, kills over 12 million people each year, most in developing countries. Air pollution kills nearly 3 million more. Heavy metals and other contaminants also cause widespread health problems.

Food supply. Will there be enough food to go around? In 64 of 105 developing countries studied by UN Food and Agriculture Organisation, the population has been growing faster than food supplies. Population pressures have degraded some 2 billion hectares of arable land—an area the size of Canada and the US.

Fresh water. The supply of freshwater is finite, but demand is soaring as population grows and use per capita rises. By 2025, when world population is projected to be 8 billion, 48 countries, containing 3 billion people will face shortages.

Coastlines and oceans. Half of all coastal ecosystems are pressured by high population densities and urban development. A tide of pollution is rising in the world's seas. Ocean fisheries are being overexploited, and fish catches are down.

Forests. Nearly half of the world's original forest cover has been lost, and each year another 16 million hectares are cut, bulldozed, or burned. Forests provide over US$400 billion to the world economy annually and are vital to maintaining healthy ecosystems. Yet, current demand for forest products may exceed the limit of sustainable consumption by 25 per cent.

Biodiversity. The earth's biological diversity is crucial to the continued vitality of agriculture and medicine—and perhaps even to life on earth itself. Yet human activities are pushing many thousands of plant and animal species into extinction. Two of every three species is estimated to be in decline.

Global climate change. The earth's surface is warming due to greenhouse gas emissions, largely from burning fossil fuels. If the global temperature rises as projected, sea levels would rise by several meters, causing widespread flooding. Global warming also could cause droughts and disrupt agriculture.

Toward a Livable Future

How people preserve or abuse the environment could largely determine whether living standards improve or deteriorate. Growing human numbers, urban expansion, and resource exploitation do not bode well for the future. Without practicing sustainable development, humanity faces a deteriorating environment and may even invite ecological disaster.

Taking action. Many steps toward sustainability can be taken today. These include using energy more efficiently; managing cities better; phasing out subsidies that encourage waste; managing water resources and protecting freshwater sources; harvesting forest products rather than destroying forests; preserving arable land and increasing food production through a second Green Revolution; managing coastal zones and ocean fisheries; protecting biodiversity hotspots; and adopting an international convention on climate change.

Stabilizing population. While population growth has slowed, the absolute number of people continues to increase by about 1 billion every 13 years. Slowing population growth would help improve living standards and would buy time to protect natural resources. In the long run, to sustain higher living standards world population size must stabilize.

18

Global Warming

Worrisome Signs

Scientists increasingly agree that the earth's atmosphere is becoming warmer. A long-term rise in the global climate could cause sea levels to rise around the world and bring a number of other adverse consequences. Reliance on fossil fuels as an energy source and the widespread destruction and burning of forests are chiefly responsible for the carbon emissions—the so-called greenhouse gases—that lie behind global warming.

One indication of global warming is that over the past 40 years the ocean surface (the top 1,000 feet) has warmed an average of half a degree Celsius. The US National Oceanic and Atmospheric Administration (NOAA) has reported that tropical waters in the Northern Hemisphere have been warming up even faster—in fact, 10 times faster than the measured global rate—because tropical oceans retain heat more readily than other areas.

Rising Sea Levels

Studies project that by 2100 the earth's surface temperature could increase between 1.0 and 3.5 degrees Celsius. If the highest projection were reached, Greenland's ice sheet probably would melt. As a consequence, the global sea level gradually would rise as much as seven meters.

Computer models project that this rise in sea level would take more than a millennium. Some climatologists, however, think that sea levels could rise much faster, printing to dramatic shrinkage of the Arctic ice cap over the past 30 years.

Even a rise of one meter in sea level—which could occur by 2080, according to the computer models—would inundate many low-lying coastal areas around the world. For instance, much of the Nile River Delta of Egypt would disappear. A one-meter rise in global sea levels also would inundate close 20 per cent of the coastline of Bangladesh and displace millions of people.

Adverse Health Effects

Rising global temperatures also would carry adverse health consequences. As temperatures warmed and episodes of droughts and floods became more frequent, the incidence of water-borne diseases and a resurgence and spread of infectious diseases carried by mosquitoes and other disease vectors probably would increase.

Warmer global temperatures also would magnify the effects of human activities on the environment, including more pollution and habitat destruction. Climate change might even cause some ecosystems to exceed critical thresholds, leading to their irreversible decline.

Growing Scientific Consensus

In 1988, to help study and focus attention on the issues, the intergovernmental panel on climate change was created under the auspices of the World Meteorological Organization and the United Nations Environment programme. The panel has involved as many as 2000 scientists from around the world. In 1996 a panel report concluded firmly that global climate change is a reality and not just a possibility. After reviewing the evidence, the panel determined that:

- evidence for the link between climate change and human activities is compelling. Already, increases in carbon dioxide and other climate-changing gases have upset the balance of the earth and its atmosphere.
- the earth's surface has become warmer; the number and severity of storms have increased; and the global sea level has risen by 10-25cm over the past century.

Because the warming trend is a global problem, solutions must be global in scope, the panel concluded.

Why is the Climate Changing

Over the last 150 years burning of fossil fuels has released some 270 billion tons of carbon into the atmosphere in the form of heat-trapping carbon dioxide gases. Since 1950 annual worldwide carbon emissions have increased fourfold, reaching 6.3 billion tons in 1997. Other emissions that contribute to climate change include methane (mainly from domestic livestock and agriculture), nitrous oxide, and chlorofluorocarbons.

Atmospheric concentrations of carbon dioxide reached 363 parts per million in 1998, the highest level since the time of massive volcanic activity over 160,000 years ago, based on examination of ice cores in Antarctica and in the Arctic. If current trends continue, atmospheric concentrations of carbon dioxide would double during this century.

About three-fourths of the huge increase in carbon emissions over the past half-century is due to increased energy consumption per capita; about one-quarter is due to population growth. Western industrialized countries account for nearly half of atmospheric carbon emissions, but developing countries are producing a growing share as industrial activity increases and populations grow. China is now the world's second largest carbon emitter, after the US.

Vanishing Carbon Sinks

The earth's forests are carbon sinks that currently soak up an estimated one-third of the carbon dioxide released into the atmosphere. When forests burn, whether naturally or when people clear the land, they not only release more carbon into the atmosphere but also diminish the amount of carbon-absorbing forest cover remaining.

Some scientists are concerned that droughts caused by global warming will increase the number of forest fires, thus contributing further to carbon emissions in the atmosphere. For instance, the six months of extensive forest fires that occurred in Asia in 1997 and 1998 released more carbon into the atmosphere than western Europe emits in a year. Burning trees for land clearance in the tropics releases about 1 billion tons of carbon into the atmosphere annually.

As more carbon fills the atmosphere, scientists worry that forests will become saturated and no longer play their role as carbon sinks. Instead, they will start to release carbon themselves.

Agriculture at Risk

Higher carbon dioxide levels in the atmosphere would extend the agricultural growing season and promote forest growth in the short-run but would have potentially negative effects on crops and forests in the long run. Because the world's grain belts would become less productive, an additional 350 million people would go hungry by the middle of this century. Major droughts have been projected for sub-Saharan Africa as climatic patterns shift, reducing rainfall and drying out soils for longer periods.

In 1999 NOAA projected that by the middle of this century soils in agricultural regions of the central US, Central Asia, and the areas surrounding the Mediterranean Sea would likely experience substantial reductions in soil moisture during the summer growing season because evaporation rates would be higher. Such reductions in soil moisture would make these areas particularly vulnerable.

Others point out that, ironically, global warming could produce colder temperatures in Northern Europe and Russia, reducing crop yields in these regions as well. This change could occur because the huge amounts of arctic freshwater from melting ice caps would make the water less dense. Such a change would interrupt the "conveyor belt" effect of the North Atlantic Drift, the ocean current that transports warm tropical water from the Gulf Stream to Scandinavia and Northern Europe.

What Can be Done?

What is the prospect for reducing emissions of carbon dioxide into the atmosphere? The United Nations framework Convention on Climate change was opened for signature at the Rio Earth summit in 1992. It was promptly signed and ratified by most low-lying island states and countries with extensive coastal areas. The Convention established a framework and a process for agreeing on specific actions later on; it asked signatory states to take preliminary action to reduce greenhouse gas emissions; and it encouraged scientific research on climate change.

19

Living with Diversity

Fishers' nets and loggers' saws may directly impoverish local ecosystems, but most biological losses have root causes far away, in long-settled urban areas and farms where diversity is seldom a concern, but where steadily rising demand for food, water, wood and other resource—and the dispersal of resulting wastes-reach far beyond the settled areas themselves. In general, these peopled landscapes have lost much of their own biological wealth, but what remains is still important to their continued functioning and livability. Reconciling farms and cities with diversity will require stopping the damage they bring to remaining natural habitats, but also beginning to halt and reverse the homogenization of these unnatural habitats.

Uniformity is not inherently undesirable. In fact, to some degree, homogeneity is the basis of all agriculture: a given type of plant is favoured and others are suppressed or eliminated. But trends in recent decades (most notably the Green Revolution and the parallel intensification of farming systems in industrial nations) have pushed uniformity to dangerous levels.

The unsustainability of modern agriculture is in part a measure of its inability to tolerate diversity. Both genetic and ecological uniformity—the sameness of fields sown horizon to horizon without interruption—demand costly and often futile reliance on chemicals to protect crops from pests or diseases that are rapidly spreading and evolving. The drive to leave no hectare unplowed worsens soil erosion, pushing tractors onto highly erodible hillsides and removing windbreaks, hedgerows and other remnant habitats.

Some of agriculture's biological impacts are obvious—the expansion of farms onto forests and wetlands, for example. While the increasing reliance on chemical inputs and machinery has reduced these impacts in some cases by decreasing the area needed to produce a given amount of food, it has worsened others.

A fundamental transition away from today's wasteful and polluting farming systems is needed to put the world's food supplies on a secure footing. Many of the reforms that will reduce farming's dependence on fossil fuel inputs and its misuse of soils and waters can also restore diversity to agricultural landscapes. Pesticides, for example, kill not only pests but other animals, such as pollinators and predators, that are beneficial to agriculture. Alternative pest control measures that lower pesticide use can also, ironically, reduce pest damage to crops by reviving the diversity of soil and insect communities, which play crucial roles in maintaining soil productivity and checking the spread of pest outbreaks.

Traditional agroecosystems are important not only because they provide sustenance to rural people and harbour valuable genetic resources, but also because they contain the seeds of a sustainable, diversity-based mode of agriculture. At varying levels, diversity is the basis of production for many peasants. Farmers often mix strains of a given crop in their fields as a hedge against the vagaries of weather. They also tend to recognize the dependence of their farms on adjacent ecological systems and to tolerate wild plants (often crop relatives whose continued interbreeding with domestic descendants contributes to genetic variety) on the outskirts of their fields.

Population growth and the expansion of large commercial farms have rendered many once-sound practices no longer viable, and traditional agriculture badly needs infusions of money and research to increase its modest yields without abandoning its stability.

Urban areas, with good reason, are considered the antithesis of natural diversity. Only the most resilient creatures (many of them regarded as weeds and pests) thrive in them, and cities' ceaseless expansion, consumption of resources, and emissions of waste threaten both farmland and wilderness almost everywhere. As with agricultural lands, the first priority for urban areas is to half their expansion onto other ecosystems and reduce the damage they export, such as the sewage poured onto coral reefs by burgeoning coastal cities throughout the tropics, or the wasteful consumption of tropical hardwoods in Japanese building construction.

But even concrete jungles can support some diversity. Landscaping of private yards and public spaces with native vegetation cannot only reduce the expense and environmental impact of watering, spraying and hauling the remains of sterile grass monocultures, but also help revive bird and other wildlife populations. Most urban areas also have water-ways running through them, or corridors of unused land such as steep ravines; if their use as waste receptacles is reduced, these can be maintained or restored as wildlife habitat.

In developing nations, especially, a surprising amount of agricultural production takes place within city limits, in home gardens. These hidden farmlands contain a great deal of genetic diversity, and their expansion could help reduce the scale and environmental impacts of commercial agriculture.

One reason that the destruction of biological diversity has gone so far without major public commitments to stopping it is that urban dwellers have little experience of the natural and even less understanding of its importance. Restoring nature where people live—reestablishing a personal link with the living world—may be necessary to save it elsewhere. For all the rational arguments favouring long-term protection of biological assets, people who have lost all direct sense of their dependence on natural systems may simply not care.

Only a growing respect for diversity for its own sake—beginning, perhaps, with a reconnection between people and nature within the urban environment—will trigger altruistic responses among those wealthy enough to have the option of considering the needs of future generations and natural communities. Although many conservation measures make economic sense, arguments of economics or self-interest will likely fail to be convincing when the contest is between a few uncharismatic species of unknown value and a major industrial project. "Human beings make sacrifies for what they love." Those who maintain strong bonds with the biological world on which they depend may be more inclined to make the hard decisions needed to protect it.

20

Ecotourism or Ecocide?

Ecotourism is the fastest growing part of the world travel business, but whether it destroys more than it protects will depend upon how it is put into practice.

For the travel and tourism industry, ecotourism is the fastest growing 'market segment', generally equated with nature tourism. Interpreted merely as a product, however, it may be ecologically based but not ecologically sound, responsible or sustainable.

To incorporate these vital characteristics ecotourism must adhere to three essential principles: The first is, perhaps, the most obvious. As an industry based on the beauty and diversity of nature, it is evident that it should not deplete or degrade those resources and thus prejudice its own future. Ecotourism must, therefore, be ecologically sound, requiring a two-way link between itself and environmental conservation.

To consider nature without recognizing the link with people will, however, compromise sustainability. It is now widely recognized that conservation cannot be divorced from development issues. The second principle is therefore that ecotourism must be responsible, paying regard to local needs and improving local welfare.

However, to be truly sustainable, ecotourism needs to fulfil the ambitions and expectations of all interests. The third principle, then, is to consider not only the interests of tourism enterprises and organizations, but also visitor satisfaction, the needs of tourists.

If ecotourism embodies these essential principles, symbiotic relationships between the varying interests should follow, with environmental protection resulting both from and in enhanced standards of living for local populations, continued profits for the tourism industry, sustained visitor attraction, and revenue for conservation. An examination of ecotourism across these dimensions highlights not only its potential but also its problems.

Local Benefits

The high ground claimed by ecotourism, in terms of its contribution to development, is that, in principle, it offers enhanced prospects for local involvement compared with conventional tourism. As well a moral obligation to incorporate the local people in projects that affect them, such incorporation has important developmental implications.

Tourism income may be captured locally through revenue sharing schemes, through entrepreneurship and labour, and through the sale of tourist merchandise. Tourism can also act as a catalyst, and even provide some of the finance, for the improvement of essential services such as clean water, sanitation, electricity supply and transport. It may also provide an incentive for improved education and skills and the potential for participation in decision making.

Local involvement also makes sense for conserving natural environments. It has been recognized that, as people realize the benefits from ecotourism, support for conservation increases. Ecotourism may also provide the incentive for the survival of a traditional culture. The cultural and the natural are often inextricably linked to form the composite attraction of a particular ecotourism destination. The terracing of the Himalayan foothills, the hot springs at Tatopani, Sikkim, are all examples of the fusion of the natural and the cultural.

Greater local involvement makes practical sense for national and local governments, agencies and operators using local labour, expertise and knowledge. Education is a two-way process, improved understanding of local circumstances is likely to increase project efficiency. Building upon local experience and traditions provides a foundation for wise and successful development, and, simultaneously, an ecotourism asset.

Introduction to indigenous uses of natural products also broadens the base of environmental interpretation. Local involvement is not without its problems, however. Revenue sharing schemes may neither benefit the most needy, nor those most adversely affected. Beneficiaries may often be passive recipients, rather than active participants. The emphasis must be on participation rather than patronization, if traditional livelihood are removed they must be replaced with others.

Local participation, however, often consists of employment rather than entrepreneurship, where constraints of costs of entry, language, education and skills operate. Furthermore, the nature of local employment tends to be low skilled, poorly paid and often seasonal. The higher status, better paid jobs, particularly managerial positions, tend to be occupied by outsiders.

Industry Profits

The integrity of the tourism 'product' is vital to the interests of tourism entrepreneurs, and all those who are associated with them, such as tourist boards, government departments, NGO's and international aid agencies. Tourism operators benefit from public support, increased credibility and demand for associated products. Sound environmental practice often makes good business sense.

There are, however, many practical and institutional obstacles to effective ecotourism management, not the least of which will be the problems of vested interests who are more concerned with short-term profits than with the long-term.

Pressure of Numbers

Another dilemma is the sheer problem of numbers. To confine attention, however, to the consideration of small-scale, more easily managed ecotourism projects, involving small specialist groups paying high prices, is to invite ecocide at higher levels.

Rapid growth rates imply inevitable change. Psychological carrying capacity (as well as other types of carrying capacity) will probably be breached and visitor satisfaction compromised. This is especially true when visitors are concentrated in space and time.

However much a principled definition of ecotourism is advocated, it must be recognized that so-called ecotourists are not an homogeneous group. The spectrum of participants embraces hard-core nature tourists through to casual day visitors. Their behaviour and consequent impact will vary accordingly. It is essential, therefore, to attempt to much numbers and types of ecotourists with destination characteristics.

Paying for Conservation

Willingness-to-pay surveys of ecotourists across the globe show a consistent response of $10 as reasonable visitor's fee. Certain unique sites, or those harbouring more charismatic species, can support higher fees. The potential revenue for conservation is therefore evident, but often not realized.

Where the fee falls below the amount visitors are willing to pay, the capability to contribute more fully towards conservation remains latent. It is also necessary to ensure that a proportion of revenues accrues locally. Percentages of revenues directed towards conservation vary between sites. Often high proportions end up in central treasuries.

The Challenge

The major role players in ecotourism all have a stake in this sustainable development. Their present and future interests are, in many ways, tied to one another. Given the multitude, and diversity, of stakeholders a completely sustainable outcome is, however, likely to remain elusive.

The grand challenge is to reconcile sometimes complementary, but often conflicting interests. The essential dilemma is to balance demands of ever-increasing 'new-tourists' escaping from the confines and pressures of urban life, and reacting against the characteristics of mass tourism with the needs of the environment, the aspirations of tourism organizations, and, most importantly, the basic needs of the local population.

Although a win-win scenario, where all interests gain, is the ideal outcome, there will often be situations where one interest may gain at the expense of another. Naticnal parks, for example, may bring benefits for conservation and for visitors, but the local population is likely to lose out if they are excluded from their traditional activities.

The situation is fraught with discontinuities. A win situation for one interest in a particular place at a specific point in time is likely to be a loss for another. It is necessary, therefore, to recognize conflicts and identify relative costs and benefits. Arriving at the most sustainable outcome is likely to involve trade-offs. It is unlikely to be optimal either environmentally or developmentally, but, in the circumstances, it will be the most feasible and most practical. And, hopefully, ecocide will be circumvented.

21

In Defence of the City Urban Development a Key for Survival

The figures sound alarming. The towns and cities in developing countries are growing faster than ever before. By the year 2000, 2.2 billion people will live in the cities of the Third World. Their numbers are expected to double by them year 2025. But many of the cities in Africa, Asia and Latin America are already bursting at the seams. Some of the so-called megacities have more than 10 or 15 million inhabitants. Many of the live in unplanned squatter settlements, without water and electricity, in an environment of squalor, poverty, crime and disease. Neverthless, the cities seem to have lost nothing of their attraction for the rural populations. Although the larger share of the population increase in the cities of developing countries is caused by the children of people already living there, the rural-urban migration continues unabated. The cities still offer better chances for employment and education, they provide a better physical infrastructure, better health facilities and a more interesting life. Miserable as conditions in the cities often appear to be, they are usually much better than those in the rural areas. It is, therefore, an illusion to believe that the growth of the cities could be checked by concentrating the development efforts on the countryside. There is no alternative to urban development in a world will soon count some 8 billion people.

Cities have always been in the vanguard of development. The ancient civilizations of Mesopotamia, Egypt, Greece and Rome were city cultures which for the first time in human development created large, well-governed states. In Europe during the Middle Ages, the creation of towns and cities offered the rural populations a chance to evade the oppression by feudal authorities and become free citizens. Local self-government in medieval towns is at the cradle of democratic development. There is a clear separation of competences between the national, [illegible] and local level of government leaving citizens an opportunity to decide on matters which directly affect their own local environment. It is worth looking at this model when discussing ways organized to improve city governance and allow for more participation of the population.

Another fact worth looking at is the size of cities in industrialized countries. Although about three-quarters of the people live in urban areas, there are only a handful of really big cities.

Of course, the growth of towns and cities in developed countries is the result of a long historical process, deeply rooted in the particular political and economic conditions of the past centuries. In developing countries today, other conditions prevail which favour the emergence of ever bigger urban conglomerations. However, governments are able thorough appropriate investments and the location of industries educational facilities, or housing policies to influence the settlement trends in their respective countries in favour of smaller cities.

One point seems cartain, though, when considering the pros and cons of city development: the severe environmental problems facing mankind today can only be solved if people live in highly concentrated settlements rather than being spread out evenly over the whole countryside. Environment-friendly mass transport, for instance, is only possible in the cities. Fossil fuel consumption which adds to the pollution of the atmosphere will be lower when people live close to their places of work. Their supply with food, water, electricity and social amenities is cheaper and uses up fewer resources when distances are short. The use of land for housing, transport, and industry is less when buildings grow in height rather than space. Even refuse disposal and wastewater management is easier to organize in a big city than in the countryside.

What is important then is not to question the validity of city development, but to make cities and towns a better place to live in. Good city governance, more involvement of the population in decision-making, more attention paid to environmental hazards caused by congestion and low safety standards are some of the demands that must be met to cope with the problems of the cities. There is no reason to bedevil the city as the most successful form of human settlement. Since the times of Babylon, it has also been a place where many different peoples and cultures meet. A generation from now, half the human population will live in cities. We should see this as a chance for human survival.

22

Measuring Population's Impact

There is no easy way to measure the overall impact of human activities on the environment. Nevertheless, several approaches have been developed as follows:

Environmental Resource Accounting:

Environmental resource accounting attempts to place an economic value on "environmental goods and services" used-natural resources that conventionally have been regarded as free and used in common. These include unpolluted freshwater, clean air, ocean life, forests, and wetlands.

Some economists argue that the value of environmental goods and services should be incorporated into estimates of gross manufactured capital, which depreciates in value overtime, environmental capital (such as forests, fisheries, and unpolluted air and water) currently is not considered to depreciate, and no charge is made against current income as it is used. A country could exhaust its mineral resources, cut down its forests, erode its soils, pollute its acquires, and hunt its wildlife and fisheries to extinction, but measured income would not be affected as these natural assets disappeared.

If natural resources were valued in the same way that manufactured assets are valued, it might help economies learn to use them more efficiently and to conserve them in other to assure continued use in the future. Such valuations also might help indicate the economic benefits of protecting the environment, as well as the ecological benefits. In other terms, instead of continuing to draw down their "environmental capital" until it is gone, economies could begin to live on its interest, maintaining the capital for use indefinitely in the future.

I = PxAxT

The equation I= PxAxT represents another effort to describe the overall impact of humanity on the environment. In the equation:

- I is environmental impact
- P is population (including size, growth, and distribution)
- A is the level of affluence (consumption per capita), and
- T is the level of technology.

Despite is limitations—for instance, inability to assign actual values to each component or the depict changes in the factors over time—the equation is valuable. In particular, it emphasizes that developing countries with large and rapidly growing populations affect the environment, even though their levels of affluence may be low, while at the same time countries in the developed world with little or no population growth have a substantial environmental impact because consumption per capita is so high.

The equation makes clear that slowing population growth is a key part of any strategy to reduce humanity's impact on the environment. For example, even if per capita resource consumption (A) declined or technologies (T) improved enough to reduce the environmental impact (I) of humanity by 10 per cent, this gain would be wiped out in less than a decade because world population (P) is growing at over 1 per cent per year since per capita consumption of resources is expected to increase as living standards rise, protecting the environment requires more efficient production technologies, less waste, and ultimately a stable would population size.

Ecological Footprints of Nations

Everybody has an impact on the Earth, because they consume the products and services of nature. Their ecological impact corresponds to the amount to nature they occupy to keep them going. In other words, we calculate the 'ecological footprints' of these countries.

Carrying Capacity

The term "carrying capacity" refers to the number of people the earth can support. Logically, population growth must stop at some point, or the earth would become overcrowded and its resources eventually would be depleted. But what is this maximum human population?

This question has been debated since 1798, when English economists Thomas Malthus predicted that population growth inevitably would outstrip the food and water supply at some point. Since the estimates of carrying capacity have varied a great deal depending on what assumptions are made out technology, consumption levels, and other factors that are not easily forecast. Some have even argued that the earth's carrying capacity may already have been exceeded in sense that the world consumed at the rate that Americans and Western Europeans consume.

While no body can know how many people the earth could support, few would want to find out the hard way—by reaching this theoretical limit. Calculate the maximum number of people who could exist on earth seems less important than determining how resources can be used wisely and managed sustainably to improve living standards without eventually destroying the natural environment that supports life itself.

23

Aquaculture a Growing Danger to Environment in India

Aquaculture has made many Indian pond owners rich, but it is now the target of escalating protests by the region's farmers, fishermen and environmentalists alarmed by its adverse ecological and social effects. Opposition to aquaculture is fast gathering momentum in shrimp growing country like India.

Thousands of farmers have had their rice fields taken over by aquaculture companies and their rice crops destroyed by seepage of salt water from nearby ponds.

A strong grassroots movement in India has developed in the eastern coastal states, where local groups have organised themselves to prevent the building of shrimp ponds and to protect themselves from attacks perpetrated by aquaculture companies. In Andhra Pradesh, villagers have attacked aquaculture farms, uprooting the pumps and breaching the fences of the ponds. The activists recently won a Supreme Court order prohibiting new aquaculture projects in three states.

In India, fishing communities have seen sharp declines in their catch due to clearing of mangroves and river pollution caused by shrimp ponds set up along the coast. Fishermen have complained of loss of income. A 2.000-hectare ell farm destroyed valuable wetlands and caused many villages to suffer water shortage due to pumping of groundwater for the farm. Several hundred farmers have also protested against the state's acquisition of their rice fields for an aquaculture project. The farmers have taken their case to court.

The problem is not confined to Indian alone. The environmental group Accion Ecologica (Ecological Action) called for a consumer boycott of cultured shrimp because the aquaculture industry had destroyed most of the mangrove forests in world's coastal regions.

Opponents of shrimp farming are also strengthening their linkup with groups in other countries. At a meeting in Madras agreed to launch an international campaign against unsustainable aquaculture. Indian organisations also set up a national people's Alliance against the Shrimp Industry.

Although the ill effects of commercial aquaculture at the grassroots level is well known, there is still a misconception that it is environment-friendly and helps augment the food supply of poor communities. This could be because small farmers and fishermen in India have been practising aquaculture for centuries to improve their living conditions. But experts stress there is a huge difference between the traditional methods and those used by commercial firms.

Traditional aquaculture in India was usually small-scale, used low inputs and relied on natural tidal action for water exchange. In some countries like India, Bangladesh and Thailand, farmers adopted a rotating system, growing rice during part of the year and cultivating shrimp and fish the rest of the year. Chemicals, antibiotics and processed feeds were not used. In this low-yield, natural method, known as "extensive aquaculture", the harvest was small but sustainable over long periods. The catch was only for family consumption or sold in local markets.

The modern method is larger in scale and intensive or semi-intensive in nature. Requiring more capital, it is owned and operated by commercial and often foreign-owned companies. In intensive aquaculture, selected species are bred using a dense stocking rate. To maintain a large population and attain higher production efficiency, artificial feed, chemical additives and antibiotics are used. This, together with excrement from the shrimp, makes the wastewater from the ponds poisonous. Overcrowding also makes the fish very prone to disease.

24
A Crucial Encounter

Genetic tests and treatments must not be allowed to create new forms of discrimination between those who, for whatever reason, can or want to take advantage of them, and those who cannot, mostly for lack of money. If a scientific discovery can form the basis of a technology, then it is highly probable that the technology will eventually be applied. Today this lesson of history is causing anxiety among politicians, scientists and public opinion concemed about the current far-reaching developments in biotechnologies.

It is now possible to penetrate to the very essence of living things as a result of spectacular scientific advances that are gradually revealing the innermost mechanisms of life. The technologies based on this field of knowledge offer humanity for the first time astonishing powers to revolutionize the process of creating and developing human beings and, ultimately, the human species. Technically speaking, these breakthroughs could lead to the revival, in even more effective guises, of eugenic practices we hoped had been buried forever. Fortunately, this nightmare scenario seems highly unlikely.

But history also shows that new technologies are rarely applied without a framework of rules and procedures designed to ensure that they are beneficially used. Human progress has always been driven by the winds of freedom, including freedom of enquiry and initiative, but human beings have always tried to head in the right direction and to respect certain limits. The biologists have done their work: they have sown the seeds of vast possibilities. Now it is up to society to make sure that only the benefits are harvested. The biotechnology revolution beckons humanity to a crucial encounter between science and ethics.

Where human reproduction is concerned, as with technology in general, we must be guided by respect for three basic and interdependent principles—dignity, freedom and solidarity.

For human dignity to be respected, each person must be regarded as unique. This position has far-reaching consequences for human procreation. First of all, it rules out cloning as a means of reproduction because this technique, which is almost upon us, involves genetically "duplicating" an existing person. More generally, predetermining the basic characteristics of a future person, notably trying to enhance their future physical or mental capacities, violates the very essence of human individuality. This kind of engineering would end up by depriving individuals of that which is theirs alone—the mysterious processes whereby their unique genetic heritage emerges and interacts in its own unique way with their environment.

Advances in prenatal scanning and testing techniques may confront parents with grave new decisions. The danger is that various kinds of pressures or even regulations will develop which only allow "genetically correct" people to be born. This would be totally unacceptable. No authority—be it political, social or economic—should be able to enact such a "genetic order", still less impose it.

So increasing emphasis must be laid on solidarity. Genetic tests and treatments must not be allowed to create new forms of discrimination between those who, for whatever reason, can or want to take advantage of them, and those who cannot mostly for lack of money.

The risk of uncontrolled, unmonitored genetic engineering increasingly looms over us. But we are starting to see the emergence of a new "responsible" form of genetic engineering in which the power of science is subjected to the power of ethics an ethics that benefits everyone, not just a few, and looks towards future generations, not just short-term interests.

25

Solutions for a Water-Short World

As populations grow and water use per person rises, demand for fresh water is soaring. Yet the supply of fresh water is finite and threatened by pollution. To avoid a crisis, many countries must conserve water, pollute less, manage supply and demand, and slow population growth.

Caught between growing demand for fresh water on one hand and limited and increasingly polluted water supplies on the other, many developing countries face difficult choices. Populations continue to grow rapidly. Yet there is no more water on earth now than there was 2,000 years ago, when the population was lass than 3 per cent of its current size. Rising demands for water for irrigated agriculture, domestic (municipal) consumption, and industry are forcing stiff competition over the allocation of scarce water resources among both areas and types of use.

Today 31 countries, accounting for under 8 per cent of the world population, face chronic freshwater shortages. By the year 2025, however, 48 countries are expected to face shortages, affecting more than 2.8 billion people—35 per cent of the world's projected population. Among countries likely to run short of water in the next 25 years are Ethiopia, India, Kenya, Nigeria, and Peru. Parts of other large countries, such as China, already face chronic water problems.

In much of the world polluted water, improper waste disposal, and poor water management cause serious public health problems. Such water-related diseases as malaria, cholera, typhoid, and schistosomiasis harm or kill millions of people every year. Overuse and pollution of water supplies also are taking a heavy toll on the natural environment and pose increasing risks for many species of life.

What Can be Done?

It may already be too late for some water-short countries with rapid population growth to avoid a crisis. Many other countries can avoid the coming crisis if appropriate policies and strategies are formulated and acted on soon. Whether water is used for agriculture, industry, or municipalities, there is much room for conservation and better management. Effective strategies must consider not only managing the water supply better but also managing demand better.

To avoid catastrophe over the long term, it also is important to act now to slow the growth in demand for fresh water by slowing population growth. Currently, in many developing countries millions of people want to plan their families and to use contraception. Family planning programmes have played an important role in assuring individual reproductive health and in reducing national fertility levels. Continuing and expanding these programmes also can help assure that population growth eventually slows to sustainable levels in relation to the supply of freshwater.

Towards a Blue Revolution

The world needs a Blue Revolution to conserve and manage freshwater supplies in the face of growing demand from population growth, irrigated agriculture, industries, and cities—just as the Green Revolution transformed agriculture in the 1960s. A Blue Revolution will require coordinated responses to problems at local, national, and international levels.

Locally led initiatives show that water can be used much more efficiently. When communities manage freshwater resources efficiently, they also manage other natural resources better, improve sanitation, and reduce disease. At the national level, especially in water-short regions with dense populations, adopting a watershed or river-basin management perspective is a needed alternative to uncoordinated water-management policies by separate jurisdictions. At the international level countries that share river basins can fashion workable policies to manage water resources more equitably. Development agencies need to focus more on assuring the supply and management of freshwater resources and on providing sanitation as part of development and public health programmes.

A water-short world is an inherently unstable world. As the next century dawns, water crises in more and more countries will present obstacles to better living standards and better health and even bring risks or outright conflict over access to scarce freshwater supplies. Finding solutions should become a high priority now.

26

Water

An Educational and Informative Approach

The most characteristic element of our planet is undoubtedly water. Indeed, more than two-thirds of the earth's surface is covered by water—the total volume representing almost 1,500 million cubic kilometers. About 94 per cent of this water is found in the oceans, almost 6 per cent is located underground and in glaciers whereas rivers, lakes, soil moisture and atmospheric vapour, which constitute the major source of drinking water, account for a mere 0.0221 per cent of the volume.

Water is indispensable for all living organisms. Life, as we know it, is impossible without water. It is present in all aspects of our life—directly or indirectly next to the air we breathe and together with the soil that we live upon, water constitutes the most important part of our environment, our most precious resource. And yet, except in the arid or semi-arid regions of the world, its value is generally overlooked until some catastrophe—natural or man-induced—forces our attention to its worth. But even so, no sooner is the situation remedied than, more often that not, we revert to our old attitude.

The reason for this sort of indifference is undoubtedly attributable to the fact that, except in exceptional circumstances, water has always been considered as a "gift of the gods", as something that human beings are as naturally entitled to as the air they breathe. Its supply, however uneven, has always seemed inexhaustible because water has a natural regenerative cycle which, until the present century, was beyond human control or interference—or even proper comprehension. But the trend of social, political and economic evolution, notably in the past 200 years, with an increase of industry, agriculture, technology, and above all, a vertiginous population growth, as led to a dramatic revision of the age-old belief that no demands made by human populations on the natural resources of the planet are in the process of setting in motion vicious circles in the environment from which it is becoming increasingly difficult to extricate ourselves, not only as concerns the present, but far more important, for the future. Thus, the overuse—or abuse—of water resources has started affecting seriously not only the water cycle but the very nature of water in such a way that, in conjunction with other abuses of the environment, the results have been climate changes, droughts, flooding, desertification on the one hand and acid rain, water pollution and eutrophication on the other.

Actually, the problem of water is to be considered less in terms of quantity—though with a steeply increasing world population making increasingly heavier demands on a fixed quantity of water, one will sooner or later be confronted with this aspect of the problem too—than in terms of proper distribution of available resources taking into account sound management, stock-age and maintenance of quality. For among the major preoccupations of humanity in the coming years, adequate supply of freshwater to the teeming populations figures in the forefront. Between 1900 and 2000 water consumption will have globally increased tenfold and though the share of agriculture, the major consumer of freshwater, is expected to drop significantly (from 90 per cent to 62 per cent approximately), that of industry and the cities will have increased enormously (approximately, from 6 per cent to 24 per cent and 3 per cent to 8 per cent respectively).

Given the current trend of societal evolution i.e. greater emphasis on industry and increasing migration towards the cities added to the global population boom, these figures are certainly cause for concern. Not only because of the damages caused to freshwater resources through the increasing use of fertilizers in the search to maximize agricultural production to cater to the increasing populations but equally because the mushrooming of industries and urban concentrations are sources of increasing water pollution. Though the industrialized nations have more than their fair share of blame in this matter insofar as the current state of water pollution goes, for the future, it is in the developing world that lies the major source of concern. Lack of resources for adequate urban planning, the increasing role of industry in the search for economic solutions added to uncontrollable population pressures are already on the way to creating an explosive situation in a great number of developing nations with the available water supply becoming more and more inadequate in terms of quantity as well as quality. And when one considers the fact that around 80 per cent of all diseases are estimated to be water related, and that by the year 2000, 51 per cent of the world population will be urban based, one can hardly be accused of exaggeration in speaking of an explosive situation.

Attacking such a vast problem is no mean task. Water being at the very source of life, water concerns every aspect of life. Thus, be it climate change, pollution, desertification, deforestation, food production..... or whatever other major environmental problem that humanity is confronted with today, water constitutes one of the prime factors. Managing our water resources with care and intelligence for the use of present and future generations is a major responsibility which has to be shared by governments and the public alike, for no sector alone can deal efficiently with so vital a problem which affects not only the present but also the future of humanity. Again, as in the case of biodiversity and climate change, the problem of water being a global problem, international cooperation is of utmost importance since activities in one part of the planet are likely to produce consequences in other regions of the world. Concerted action by the international community alone is capable of dealing effectively with a problem of such far-reaching consequences.

If our planet is to be saved from disaster—for in jeopardising our water resources we are guilty of nothing less than condemning life itself on our planet—we have to work for sustainable results: short-term plans for the present which will dovetail into medium-term ones for the coming generations without compromising the possibility, at the same time, of careful long-term planning to guarantee the future of the planet. In this, the part of environmental education and information of the people is fundamental. No strategy, no policy, no plan—be it ever so well prepared and implemented—can hope to succeed without the active and effective participation of the main actors-the people who must be properly educated and informed. For this age-old techniques, beliefs—mentalities must be brought in line with present-day realities. People have to learn to think differently in order to veer from a course which, however right in the past, has been shown to be less than adequate for present conditions—and catastrophic for the future—and must therefore needs to be altered.

Changing mentalities is neither an easy nor a rapid process. It is difficult to go back upon the accumulated experience of generations—even in the face of stark realities and scientific evidence. Moreover, when dealing with such global and fundamental issues as water, where even "scientific evidence" tends to be stated in tentative terms, the task becomes more onerous. Add to this the fact that the problem presents itself most presently in developing countries which are equally subject to enormous economic pressures which tend to reduce the scope of possible solutions. We are thus faced with the enormous task of trying to change attitudes, values, mentalities of populations whose geographical, socio-cultural and economic conditions have already fashioned priorities other than those that would precisely permit them to overcome their difficulties in a sustainable manner. In other words, of persuading people to abandon traditional short-term strategies in favour of perhaps more unattractive but eventually sustainable, long-term practices.

A veritable Herculean labour—which can only be accomplished through information and education. And in particular, through environmental education and information whose avowed aim is precisely to develop the understanding, knowledge, skills and motivations leading to the acquisition of attitudes, values and mentalities which are necessary to deal effectively with environmental issues and problems. Sound and systematic environmental education of the people associated with concerted local, national and international action, is the only means to finding a sustainable solution to this problem. The ground has to be paved through adequate information on the subject followed by educative processes adapted to specific local conditions. For a uniform education, whether formal or non-formal, might perhaps do more harm than good as its rejection, due to its unsuitability in the light of local customs, beliefs, traditions.., might only serve to reinforce the very attitudes that [illegible]s to change. In each region, each country, each locality the educative processes must correspond to the socio-cultural, historical, economic conditions of the people. Only then can we hope to arrive at the change in mentalities around the planet which, coupled with consistent, parallel support from national and international institutions, will lead to the safeguard of what is perhaps our most precious resource—water.

27

Big-Dam Construction Is on the Rise

As the era of big dams fades in North America, construction is increasing in Asia, fueled by growing demand for electricity and irrigation. Worldwide, the number of dams under construction in 1998 (the latest year for which global data are available) rose by more than 9 per cent, to at least 1,240, after a much smaller increase in 1992. The accompanying chart shows that China accounts for more than one-fourth of the big dams under construction, while China, Japan, South Korea and India together account for more than half. The increases follow a decline in the 1980s, when construction world wide averaged less than half that of the preceding 25 years.

Data for the early 1990s, though incomplete, indicate a shift toward larger dams, with concomitant greater social and economic costs. In 1992, 60 per cent of the dams being built were more than 30 meters high compared with only 21 per cent of existing dams in 1986. Construction of dams higher than 100 meters rose by some 27 per cent between 1991 and 1993, half of these large structures were built by just three countries—Japan, China, and Turkey.

Dams are a symbol of modernity and a source of national prestige, partly because they are a multipurpose tool of development. They generated more than 18 per cent of the World's electricity in 1998, and reservoir water irrigates millions of hectares of land around the world, raising agricultural yields two to three times over those of dry land farming, and in many cases bringing agriculture to regions that could not otherwise support it. Areas with a high percentage of irrigated agriculture,

Leading Builders of Big Dams
(Higher than 10 meters)

Country	*New Starts (Number)*	*Under Construction (Number)*
China	82	311
Turkey	84	190
Japan	11	140
South Korea	2	125
India	48	76
United States	30	55
Spain	16	53
Romania	0	39
Italy	0	37
Tunisia	16	28
Algeria	6	27
Iran	1	76
Thailand	7	17
Greece	3	14
France	8	12
Brazil	4	12

such as China, India, and the western United States, rely heavily on reservoirs for water. Irrigation is increasingly listed as the primary purpose of new dams, possibly reflecting growing food needs and water scarcity in many developing countries.

Dam reservoirs also protect societies against drought. Water in the world's reservoirs effectively increases the normal supply from rivers by some 30 per cent. This insurance can be invaluable: Egyptians maintain that the Aswan Dam saved Egypt from catastrophe during the drought of 1979-80.

In the past two decades, however, as it became clear that dams have serious costs as well as benefits, dam construction began to decline. Loss of land to reservoirs can be significant. The Narmada Sagar Project in India is expected to drive 31 species of plant to extinction because of habitat changes caused by reservoir creation. Decomposing plant life at reseroir bottoms can release as much methane and carbon dioxide both greenhouse gasses—as a coalfired plant with the same electricity-generating capacity, according to the Freshwater Institute in Canada.

Dams also restrict stream flow with often disastrous biological and economic consequences. Populations and fisheries have been nearly eliminated in many places worldwide, as on the Columbia River in the United States, where once plentiful salmon are now endangered species, and valuable fisheries have collapsed. Changes in temperature and water alter habitats of fish and other wildlife. Sediment that would normally enrich flood plain soils gets trapped behind dams. This last problem, siltation, can shorten a dam's useful life; in an extreme case, the turbines of China's Sanmenxia Dam were shut down in 1964 after only four years of operation when the dam's reservoir filled with sediment.

By restricting stream flows, dams can also contribute to the decline of local economies. One example is Indian diversion of the Ganges river before it reaches Bangladesh, which has caused $25 million in losses for an agricultural project in northwest Bangladesh, and increased salination of fishing areas at the Bay of Bengal, the river's natural outlet.

Dams exact a high human toll as well. Millions of people were resettled in the past half century to make room for dam reservoirs. China's Three Gorges Dam alone may will displace more than one million people. Those affected have had little say in their resettlement, and most displaced citizens wind up in worse situations than before their move.

Dams can increase health risks. Waterborne disease, including malaria and river blindness, have been introduced and spread as reservoirs are created. Large-scale water projects were a major contributor to the 75 per cent global increase in cases of schistosomiasis, sometimes fatal disease.

Advocates of dams acknowledge these shortcomings, but argue that these effects can be mitigated, or that the alternatives to dams would be worse. Others argue that evaluation of dams involves widely differing environmental, social, and political contexts, and a project-by-project analysis of each dam is needed.

The World Bank—long a strong supporter of big dams—appears to have slowed its involvement. The Bank was involved with an average of 18 dam projects a year between 1980 and 1985, but with. only 6 a year between 1986 and 1993. In fact, the Bank established a commission in 1994 to hear complaints from parties affected by proposed dam projects, the first of whom were people affected by the Arun III dam in Nepal.

The future of dams is unclear. Proponents note that only a fraction of the technically usable hydropower potential in developing countries has been harnessed. But the discussion has moved beyond technical considerations. Wideranging development, conservation, energy use and population considerations are or should be part of every proposed dam's calculus. It remains to be seen whether current and future big dams can meet the increasingly stringent environmental and social standards expected of them, and whether development needs can be met in other ways.

28

A Breakthrough in the Evolution of Large Dams?

Back to the Negotiating Table

"The problem is not the dams. It is the hunger. It is the thirst. It is the darkness in a township". With these plain words, former South African President Nelson Mandela summed up the World Commission on Dams (WCD's) motives in a speech at the presentation praising its work. His position is similar to that of the many other representatives of the South who were present: he demands a right to development. But Ms Medha Patkar, of India, a WCD Commission member and founder of the Struggle to Save the Narmada River (Narmada Bachao Andolan) anti-dam movement, takes a contrary view. "The problems of the dams are only a symptom of the larger failure of the unjust and destructive dominant development model," she says. We need to challenge "the forces that lead to the marginalisation of a majority through the imposition of unjust technologies like large dams".

Stragnation in Dam Building

The WCD is a unique experiment in reaching consensus. It began in April 1997 when with the support of the World Bank and the World Conversation Union (IUCN), 39 representatives of diverse interests met at a workshop in Gland, Switzerland. At this point the various positions of the participants from governments, the private sector, international financial institutions, civil society organisations and affected people were cast in stone. The Manibeli Declaration in June 1994 of 326 activist groups from 44 countries had called for an immediate moratorium on world Bank funded large dams until a comprehensive, independent review of all Bank funded projects had been conducted. International financial institutions were actually on longer able to fund further large dams in the face of public criticism. Enervated by steadily growing protests despite continual tightening of social and environmental standards, the institutions' representatives wearily likened the dispute to a football match in which somebody kept moving the goalposts. But one proposal to emerge from the meeting in Gland was for all parties to work together in establishing the World Commission on Dams.

The WCD began its work in May 1998 under the chairmanship of Prof. Kader Asmal, then South Africa's Minister of Water Affairs and Forestry. Its 12 members were chosen to reflect regional diversity, expertise and stakeholder perspectives. But the Commission ran the risk of failure right from the start due to their confrontational attitudes. The members, who spent 2 years 6 months jointly organising hearings, consultations and case studies and analysing more than 100 existing large dams, could not be more disparate.

The Current Situation

A large dam is a dam with the height of 15m or more from the foundation. If dams are 5-15 metres high and have a reservoir volume of more than three million cubic metres, they are also classified as large dams. Using this definition, there are more than 45,000 large dams around the world, almost half of them in China. They were built in the 20th century to meet the constantly growing demand for water and electricity. On a global scale, hydropower dams account for about 20 per cent of electricity generated, and in 24 countries, including Brazil, Democratic Republic of Congo, Zambia and Norway, hydropower covers more than 90 per cent of national electricity supply needs. Half the world's large dams were built solely or mainly for irrigation. Between 12 per cent and 16 per cent of world food production is based on dams, and as reservoirs they provide protection against floods.

Unfortunately, this impressive balance is counteracted by comparably significant problems. Construction of large dams is a major intervention in the ecosystem of rivers and the lives of many people. The WCD estimates that some 40-80 million people, mostly indigenous peoples, have been displaced by reservoirs worldwide and robbed of their livelihoods from fishing or farming. Serious conflicts are simmering between neighbouring countries because dams have turned off the water supply for downstream states.

The late Indian Prime Minister Jawaharlal Nehru once said: "Dams are India's new temples." Right up to the 1970s, large dams were seen as the synonym for development and economic progress. Dam-building reached its peak between 1970 and 1980, when an average of two to three new large dams per day were commissioned. But a considerable number of the dams analysed by the WCD have fallen short of their technical and economic objectives. Construction cost overruns averaged 56 per cent. Many dams have had negative ecological impacts, and the disadvantages for people living downstream were mostly not taken into account. The planning of dams did not examine sufficiently possible alternatives for meeting power and water needs. There were hardly any retrospective evaluations of dam projects.

A New Framework for Decision-making?

Despite this sobering stocktaking, the WCD arrives at an astonishingly simple finding: dams are primarily a means to an end. Their task is to improve the well-being of the people on a sustainable basis. This improvement should be economically acceptable, socially just and environmentally sound. If this goal can be achieved by a dam, its construction should be supported. Where alternative options offer a better solution they should be the preferred choice.

The Commission based its work on a set of five crore values for future decision-making: equity, efficiency, participatory decision-making, sustainability, and accountability. With regard to legal aspects and the extent of the potential risks for those involved, the WCD proposes development of an approach based on recognising rights and assessing risks. All risk-bearers should have a place at the negotiating table.

The WCD also recommends seven strategic priorities for decision-making: gaining public acceptance; comprehensive options assessment; reviewing exiting dams; sustaining rivers and livelihoods; recognising entitlements and sharing benefits; ensuring compliance; and sharing rivers for peace, development and security. These priorities are reinforced by 23 practical criteria and guidelines which can be adopted, adapted and applied by all actors involved in the dam controversy, For instance, the WCD suggests analysing points at issue together with the people affected by existing dams and developing joint proposals for solutions. People affected by new dam projects should be among their favoured beneficiaries, and their claims should be made legally binding.

The report offers a comprehensive compilation of knowledge, which previously was limited to individual case studies or a narrow specialist context, on the social, economic, technological and ecological problems and impacts of large dams. That makes the report a central reference which helps greatly in bringing objectivity into the debate. Provision of an analytical framework and strategic options is certainly an important step. But with regard to its task of developing internationally valid criteria and guidelines for the planning, design, appraisal, building operation, monitoring and shutdown of dams, the report remains very general. What will be decisive here will be to practice with the relevant actors the method and content of the suggested mediation process on the basis of specific cases.

Deeds Must Follow Words

The WCD's work must now be made useable for the private sector, civil society and development purpose. What is required is a discussion process involving all major actors. The objective of this process must be the development of practical and effective guidelines in addition to the current standards.

The WCD calls on bilateral development organisations and multilateral development banks to support only dam projects that have resulted from an open process of examining various options. The parties should observe the WCD guidelines. Measures to save water and power should be examined and, if applicable, be promoted.

Private sector companies should publish guidelines on corporate behaviour and acknowledge the WCD principles, criteria and guidelines. Further, the private sector should draw up and implement voluntary codes of conduct, management systems and certification procedures, such as the internationally recognised standard for environmental management (ISO 14001). The OECD's Anti-Corruption Agreement should be observed, and declarations of honesty incorporated in contracts. Business associations should develop processes to monitor compliance with the WCD guidelines.

NGOs should primarily check compliance with agreements and assist aggrieved parties to seek compensation. They should also assist in identifying relevant stakeholders for dam projects, using the rights and risks approach. Finally the NGOs should build up support networks and partnerships between them.

Importance of Information

But do these noble proposals provide the whole answer? The WCD process is based on the opportunities for personal development of every individual in an open society. But it is not enough to build on the negotiating abilities of the potentially affected alone because only specialists can anticipate the complex impacts of dams. Therefore participation presupposes that the mediation process contains a substantial informative component.

Relying solely on a mediation process is also not sufficient in providing for social impacts. There is no generally recognised method to determine the value of 'goods' in a subsistence economy. Without objective criteria, the moral demands on both sides are extremely high. Strategically-motivated behaviour will then not be prevented even if all participants agree readily that the subjective standard of living of people affected by a dam should be improved or at least maintained.

Mediation processes make sense only if agreements are observed. According to the WCD analyses, lack of compliance with agreements is the main cause of the negative social impacts of dams. In the case of dam projects co-financed by international donors, disbursements of funding instalments could depend upon independent evaluations that confirmed compliance. Ensuring compliance is much more difficult in the case of projects financed by the private sector. Because there is no independent institution that could assume the role of arbitrator, it must be in the business world's own interests to act responsibly in ecological and social terms. To be credible, it must provide transparency and independent certifiers.

Due to the opposing interests involved, no one should expect reaching consensus to be easy. But the example set by the WCD is not the only reason for hope. Taking a closer look at it, the model offers significant advantages for all participants. Partner countries and development organisations wish to continue to use the potential for development which dams will also offer in the future. The private sector will also continue to build and operate dams, and for that they need planning certainly. The advantages for NGOs and the people affected are also obvious.

29

The Coming Water Crisis

Freshwater is emerging as one of the most critical natural resource issues facing humanity. The world's population is expanding rapidly. Yet there is no more freshwater on earth now than there was 2,000 years ago, when the population was less than 3 per cent of its current size.

Water is, literally, the source of life on earth. The human body is 70 per cent water. People begin to feel thirst after a loss of only 1 per cent of bodily fluids and risk death if fluid loss nears 10 per cent Human beings can survive for only a few days without freshwater. Yet, in a growing number of places people are withdrawing water from rivers, lakes, and underground sources faster than they can be recharged—"unsustainably mining what was once a renewable resource," as one researcher puts it. Currently, 31 countries—mostly in Africa and the Near East—face water stress or water scarcity.

Population growth alone will push an estimated 17 more countries, with a projected population of 2.1 billion, into these water-short categories within the next 30 years. By the year 2025, 48 countries, with more than 2.8 billion people—35 per cent of the projected global population in 2025—will be affected by water stress or scarcity. Another nine countries, including China and Pakistan, will be approaching water stress.

Beyond the impact of population growth itself, the demand for freshwater has been rising in response to industrial development, increased reliance on irrigated agriculture, massive urbanization, and rising living standards. In this century, while world population has tripled, water withdrawals have increased by over six times. Since 1940 annual global water withdrawals have increased by an average of 2.5 per cent to 3 per cent a year compared with annual population growth of 1.5 per cent to 2 per cent. In developing countries over the past decade water withdrawals have been increasing by 4 per cent to 8 per cent a year.

Moreover, the supply of freshwater available to humanity is shrinking, in effect, because many freshwater resources have become increasingly polluted. In some countries lakes and rivers have become receptacles for a vile assortment of wastes, including untreated or partially treated municipal sewage, toxic industrial effluents, and harmful chemicals leached into surface and ground waters from agricultural activities.

Caught between finite and increasingly polluted water supplies on one hand and rapidly rising demand from population growth and development on the other, many developing countries face uneasy choices. The lack of freshwater is likely to be one of the major factors limiting economic development in the decades to come.

Slowing Demand, Conserving Supplies

To avoid water crisis, particularly in water-short countries with rapid population growth in demand for water by managing the resource better, while at the same time slowing population growth as soon as possible. Family planning programmes play an important role not only for individual reproductive health but also for sustainability of the use of freshwater and other natural resources in relation to population size.

As population grows, so does demand for freshwater for food production, household (municipal) consumption, and industrial uses. The availability of freshwater limits the number of people that an area can support and affects standards of living. In turn, population growth and density typically affect the availability and quality of water resources in an area, as people attempt to assure their water supply by digging wells, constructing reservoirs and dams, and diverting the flow of rivers. If needs consistently outpace available supplies, at some point overuse of water leads to the depletion of surface and groundwater resources, triggering chronic water shortages.

Scarce and unclean water supplies are critical public health problems in much of the world. Polluted water, water shortages, and insanitary living conditions kill over 12 million people a year.

Competition for freshwater supplies breeds social and political tensions. River basins and other water bodies do not respect national borders. For example, one country's use of upstream water often subtracts from the supply available for use by downstream countries. As the 21st century dawns, there is a growing risk that wars will be fought over access to freshwater supplies.

If a crisis is to be averted, the world's overuse and misuse of freshwater must end as soon as possible. We cannot afford to keep wasting and fouling our precious supplies of freshwater. Increasingly, human activities are altering the flow of water and drawing down freshwater supplied faster than they can be replenished. Throughout the world enormous amounts of water are wasted due to inappropriate agricultural subsidies, inefficient irrigation systems, leaky municipal pipes, improper pricing of municipal water, poor watershed management, and other imprudent practices. It is time for widespread conservation measures, effective water management policies, and growing attention to assuring freshwater supplies and decent sanitation as part of development and public health projects.

30

Water Facts and Findings on Large Dams

Water Facts

Today, around 3800 km^3 of fresh water is withdrawn annually from the world's lakes, rivers and aquifers. This is twice the volume extracted 50 years ago. World population has passed 6 billion. Projections say that it will reach a peak of between 7.3 billion and 10.7 billion around 2050 before total population begins to stabilise or fall.

Fifty litres per person per day (or just over 18.25m^3 a year) covers basic human water requirements for drinking, sanitation, bathing and food preparation. In 1990, over a billion people had access to less than 50 litres of water a day. Agriculture accounts for about 67 per cent of withdrawals, industry uses 19 per cent and municipal and domestic uses account for 9 per cent.

One-third of the countries in water stressed regions of the world are expected to face severe water shortages this century. By 5025 there will be approximately 6.5 times as many people—a total of 3.5 billion living in water-stressed countries. By the end of the 20th century, there were over 45,000 large dams in over 150 countries. The average large dam today is about 35 years old. Since average construction periods generally range from 5 to 10 years, this indicates a worldwide annual average of some 160 to 320 new large dams per year.

During the 1990s, an estimated $ 32-46 billion was spent annually on large dams, four-fifths of it in developing countries. Of the $ 22-31 billion invested in dams each year in development countries. Of the $ 22-31 billion invested in dams each year in developing counties, about four-fifths was financed directly by the public sector. About one-fifth of the world's agricultural land is irrigated, and irrigated agriculture accounts for about 40 per cent of the world's agricultural production.

Half the world's large dams were built exclusively or primarily for irrigation, and an estimated 30 to 40 per cent of the 271 million hectares of irrigated lands worldwide rely on dams. Dams are estimated to contribute to 12-16 per cent of world food production. Hydropower currently provides 19 per cent of the world's total electricity supply, and is used in over 150 countries with 24 of these countries depending on if for 90 per cent of their supply.

Floods affected the lives on average, of 65 million people between 1972 and 1996, more than any other type of disaster, including war, drought and famine. There are 261 water sheets that cross the political boundaries two are more countries. A number of key international rivers lack a basin-wide agreement that defines a process for establishing equitable water use between riparian States.

Cost Effectiveness

Cost performance date confirms that large dam projects often incur substantial capital cost overruns. The average overrun was half again as much as the projected cost. The bulk of hydropower projects have delivered power within a close range of pre-project targets but with an overall tendency to fall short of targets.

At current rates, water fees are rarely sufficient to recover both capital and recurrent costs for water supply systems in many developing countries. Growing concern over the cost and effectiveness of large dams and related structural measures as long-term responses to floods has led to support for integrated flood management as opposed to flood control.

Multi-purpose schemes are inherently more complex, and many experience operational conflicts that contribute to under-performance on financial and economic targets. Substantive evaluations of project performance are few in number, narrow in scope, and poorly integrated across impact categories and scales.

Ecological Costs

Dams, inter basin transfers, and water withdrawals for irrigation have fragmented 60 per cent of the world's rivers. As a physical barrier the dam disrupts the movement in upstream and downstream species composition and even species loss. In Africa, the changed hydrological regime of rivers has adversely affected floodplain agriculture, fisheries, pasture and forests that constituted the organising element of community livelihood and culture.

Problems may be magnified as more large dams are added to a river system, resulting in an increased and cumulative loss of natural resources, habitat quality, environmental sustainability and ecosystem integrity. Good site selection, such as not building large dams on the main-stem of a river system, and better dam design also played significant roles in avoiding or minimising impacts. The economic appraisal techniques such as risk and distributional analysis were still mandated for only 20 per cent of large dam projects even in the 1990s.

Social Costs

At the planning and design stage, an important social impact is the delay between the decision to build a dam and the onset of construction. This can result in communities living for decades starved of development and welfare investments.

The overall global level of physical displacement could range from 40 to 80 million. In India and China together, large dams could have displaced between 26-58 million. People between 1950 and 1990. Little or no meaningful participation of affected people in the planning and implementation of dam projects-including resettlement and rehabilitation has taken place.

Empowering people, particularly the economically and socially marginalised by respecting their rights and ensuring that resettlement with development becomes a prócess governed by negotiated agreements is critical to positive resettlement and rehabilitation.

Poor accounting in economic terms for the social and environmental costs and benefits of large dams implies that the true economic efficiency and profitability of these schemes remains largely unknown.

The direct adverse impacts of dams have fallen disproportionately on rural dwellers, subsistence farmers, indigenous people, ethnic minorities, and women. Where costs and benefits accrue to different groups, the standard procedures for adding up and discounting the expected costs and benefits do not provide an appropriate measure of changes in social welfare.

Financing

Almost 2 billion people, both urban and rural poor, have no access to electricity at all. Most efficiency measures and technologies are cost-effective at today's electricity prices and the use of full environmental and social costing of electricity supply options makes them even more so. Among advanced technologies in research and development, micro turbines and fuel cell show the greatest near and mid-term promise.

Total financing for large dams from multi-lateral and bi-lateral development banks comes to more than $4 billion annually at the peak of lending during 1975-84. Although the proportion of investment in dams directly financed by bilaterals and multilaterals was perhaps less than 15 per cent. The total investment in dams by the multilaterals and bilaterals since 1950 is approximately $ 125 billion.

31

A Rare and Precious Resource

Fresh water is a scarce commodity. Since it's impossible to increase supply, demand and waste must be reduced. But how?

Water is a bond between human beings and nature. It is ever-present in our daily lives and in our imaginations. Since the beginning of time, it has shaped extraordinary social institutions, and access to it has provoked many conflicts.

But most of the world's people, who have never gone short of water, take its availability for granted. Industrialists, farmers and ordinary consumers blithely go on wasting it. These days, though, supplies are diminishing while demand is soaring. Everyone knows that the time has come for attitudes to change.

Few people are aware of the true extent of fresh water scarcity. Many are fooled by the huge expanses of blue that feature on maps of the world. They do not know that 97.5 per cent of the planet's water is salty—and that most of the world's fresh water—the remaining 2.5 per cent—it unusable: Seventy per cent of it is frozen in the icecaps of Antarctica and Greenland and almost all the rest exists in the form of soil humidity or in water tables which are too deep to be tapped. In all, barely one per cent of fresh water—0.007 per cent of all the water in the world, is easily accessible.

Over the past century, population growth and human activity have caused this precious resource to dwindle. Between 1900 and 1995, world demand for water increased more than sixfold—compared with a threefold increase in world population. The ratio between the stock of fresh water and world population seems to show that in overall terms there is enough water to go round. But in the most vulnerable regions, an estimated 460 million people (8 per cent of the world's population) are short of water, and another quarter of the planet's inhabitants are heading for the same fate. Experts say that if nothing is done, two-thirds of humanity will suffer from a moderate to severe lack of water by the year 2025.

Inequalities in the availability of water—sometimes even within a single country—are reflected in huge differences in consumption levels.

Scarcity is just one part of the problem. Water quality is also declining alarmingly. In some areas, contamination levels are so high that water can no longer be used even for industrial purposes. There are many reasons for this—untreated sewage, chemical waste, fuel leakages, dumped garbage, contamination of soil by chemicals used by farmers. The worldwide extent of such pollution is hard to assess because data are lacking for several countries. But some figures give an idea of the problem. It is thought for example that 90 per cent of waste water in developing countries is released without any kind of treatment.

Things are especially bad in cities, where water demand is exploding. For the first time in human history, there will soon be more people living in cities than in the countryside and so water consumption will continue to increase. Soaring urbanization will sharpen the rivalry between the different kinds of water users.

Curbing the Explosion in Demand

Today, farming uses 69 per cent of the water consumed in the world, industry 23 per cent and households 8 per cent. In developing countries, agriculture uses as much as 80 per cent. The needs of city-dwellers, industry and tourists are expected to increase rapidly, at least as much as the need to produce more farm products to feed the planet. The problem of increasing water supply has long been seen as a technical one, calling for technical solutions such as building more dams and desalination plants. Wild ideas towing chunks of icebergs from the poles have even been mooted.

But today, technical solutions are reaching their limits. Economic and socio-ecological arguments are levelled against building new dams, for example: dams are costing more and more because the best sites have already been used, and they take millions of people out of their environment and upset ecosystems. As a result, twice as many dams were built on average between 1951 and 1977 than during the past decade.

Hydrologists and engineers have less and less room for manoeuvre, but a new consensus with new actors is taking shape. Since supply can no longer be expanded—or only at prohibitive cost for many countries—the explosion in demand must be curbed along with wasteful practices. An estimated 60 per cent of the water used in irrigation is lot through inefficient systems, for example.

Economists have plunged into the debate on water and made quite a few waves. To obtain "rational use" of water, i.e. avoiding waste and maintaining quality, they say consumers must be made to pay for it. Out of the question, replay those in favour of free water, which some cultures regard as "a gift from heaven". And what about the poor, ask the champions of human rights and the right to water? Other important and prickly questions being asked by decision-makers are how to calculate the "real price" of water and who should organize its sale.

The State as Mediator

The principle of free water is being challenged. For many people, water has become a commodity to be bought and sold. But management of this shared resource cannot' be left exclusively to market forces. Many elements of civil society-NGO's, researchers, community' groups—are campaigning for the cultural and social aspects for water management to be taken into account.

Even the world Bank, the main advocate of water privatization, is cautious on this point. It recognizes the value of the partnerships between the public and private sectors which have sprung up in recent years. Only the state seems to be in a position to ensure that practices are fair and to mediate between to parties involved—consumer groups, private firms and public bodies. At any rate, water regulation and management systems need to be based on other than purely financial criteria. If they aren't, hundreds of millions of people will have no access to it.

32

Fresh Water and the Environment

It is widely recognized that water is going to be one of the major issues confronting humanity at the turn of the century and beyond. We are facing a crisis as regards the quantity and quality of water supply, but we have yet to experience the full social and political impact of that crisis. The escalation in the population and the quest for continued development is leading to conflicting pressures on water resources. Such resources are the ultimate recipient of pollution from various socio-economic activities associated with urbanization, agriculture, mining and clearing of native vegetation. Pollution originating from human waste, especially where appropriate sanitation facilities are not available, or are located too close to water supply sources affects both surface water and ground water.

This makes water supply and health perhaps the most important issue for the larger proportion of the global population. Paradoxically, the demands for "sustainable management" and increasing global population require more potable water from a declining available potable water base.

It is universally accepted that proper water administration is a critical component of sustainable development—that is, development that meets the needs of both present and future generations. Indeed, water is an essential factor in a large number of productive activities, of which one of the most important is the production of food by irrigation. This activity accounts for two-thirds of the water resources used by humanity. A supply of drinking water and sanitation in urban centers are crucial for preserving human health.

For some decades it has been known that the misuse of water resources is responsible for many important environment problems. For example, in many industrialized cities both surface water and ground water are seriously contaminated. This deterioration is a consequence of a range of human activities, sometimes in isolation, others over a large area or a long period of time. Among examples of the latter is modern agriculture, whether it uses irrigation or not, as a result of the intensive use made of mineral fertilizers and pesticides.

Water Shortage: Exaggeration, Reality or Bad Management?

Some of these problems have made news and have created the impression that water shortage will be one of humanity's big problems in the coming decades. Sometimes this feeling is due to genuinely manipulative publicity campaigns to justify the setting in motion of hydraulic megaprojects which basically benefit a few large construction companies. The truth is that except for a handful of very specific cases, no problems of water shortage are to be found almost anywhere On the other hand, cases of bad water management are not rare at all.

Basic Principles for Good Water Management

Good management of water resources—and of almost all other natural resources—must be based on the principles of solidarity, "subsidiarity" and participation. The physical reality requires that these resources be considered a common heritage of humanity both now and in the future. By "subsidiarity" we mean that water management should be as decentralized as possible: what one person or any minor social group can do should not be done by a higher authority. For example, what local government can do should not be done by a regional, state or central government. Participation consists in water users playing as large a part as possible in decisions affecting water, in keeping with each state's or country's social and cultural structure. Obviously this participation calls for a certain cultural and technical knowledge—a hydrological education—on part of the those users.

The need for participation by users is even greater in the exploitation of groundwater. In this case, users tend to extract water independently of one another. They often fail to realize, until there is a serious economic or environmental impact, that their pumping affects other people who rely on the same water supply as has happened.

Water shortage is rarely a serious problem: in fact, in some cases the problem is exaggerated to justify the construction of large works using taxpayers' money. On the other hand, the contamination of surface and groundwater tends to be a problem which rarely receives adequate treatment. Successful water management should be based on three basic principles: solidarity, subsidiarity and participation. The specific way in which these principles are applied will vary from one state or country to another, but the effectiveness of water management will depend in large measure on the hydrological education of the general public.

The universal way of obtaining freshwater is from rain. River systems are the results of the excess water that falls on dry land in the form of rain. On the one hand, rainwater penetrates the permeable soils, saturates them and accumulates to form groundwater reservoirs, or aquifers, which can come to the surface in the form of springs. On the other hand, the water is absorbed by vegetation, which uses it for pumping minerals and them evaporates it by transpiration. Some rainwater is lost because it evaporates immediately on falling on impermeable surfaces like the asphalt of roads and cities. Running water courses finally flow over saturated soils, shaping the complex systems of the watersheds or river basins.

Since each basin's natural system has developed gradually and has grown up according to the yearly distribution and fluctuations of rainfall, we have to appreciate that any large-scale project for redistributing water by means of pipes, as if it were gas or electricity, is a journey into the unknown. This is because it destroys the results of the work of shaping the climate, however transitory it might be.

Variable Volumes

All water supplies are of variable volume. Both the discharge of rivers and the level of lakes and aquifers depend on rainfall. As these resources are components of a larger system, the river basin, a reasonable policy would be to manage water resources according to the characteristics of each basin. This would require, first of all, a proper understanding of the system so as to adapt use and consumption to the existing supply. Conserving river systems as much as possible in their natural state is the best guarantee for the preservation of the landscape and of a constant supply. Groundwater reservoirs aren't canals, but are more like lakes, so that pollution leads to the build-up of a debt which is paid in years to come.

Consumption

Water consumption has increased in recent years as a result of not only population growth but also an increase in living standards. In the rural areas the introduction of new farming methods, the spread of irrigation and the excessive use of fertilizers and pesticides causes very high consumption—it is estimated that more than two-thirds of water consumption is used in irrigating. Agricultural pollution also endangers both surface water and aquifers, which receive water full of chemical products. Many cases of eutrophication, the enrichment of water by nutrients that accelerate the growth of algae, derive from the run-off of fertilizers. The practice of intensive stock-raising on farms with large numbers of animals also brings about these problems of overconsumption and pollution. Cleaning the stockyards requires large amounts of water which is then released into the environment with high concentrations of nitrogen.

As for industries, they have in the past taken little care over water consumption and dumping, and in many areas the need for proper attention comes as something new. The best thing would be to make industry take its water at a point down-river from where it returns it or, better still, generalize the use of closed circuit systems based on the constant recycling and reusing of the same water.

As regards human consumption, the general attitude to cleanliness is based on diluting pollutants. One example is the success of the use of the Water Closet which involves diluting a few decilitres of urine in 10 or more litres of drinking water: quite a record in wastefulness.

Another aspect to be considered is the different quality of the water that falls on well formed soils from the water that falls on roads, cities, airports, suburbs and built-up areas and whose composition is less stable and "worse" than that resulting from a more uniform interaction with mature soils. Remember that streets, roofs, communication routes, airports and built-up areas already cover a high proportion of the earth's land area and are still on the increase.

Purification techniques should be based especially on the natural processes that include biological activity. Otherwise, for example, if physico-chemical methods are used—there can be side-effects such as an excess of mud or sediments. The strategy to follow is to optimize operations in our use of water according to the discharge and to the distribution of contamination. A system in the form of a conduit or channel, such as a river, can respond relatively quickly. On the other hand, lakes and dams can only do so up to a point, because they show more inertia and irreversibility and take longer to clean.

Large lakes, not to mention the sea, might seem a good place to dump contaminating refuse, but they can't then be cleaned. This is the price we pass on the future generations: a comfortable attitude, but an unacceptable one.

33

Pro-Poor Tourism

Opportunities for Sustainable Local Development

Tourism is the world's largest industry, with over 10 per cent of GDP globally directly related to tourism activities. Rising standards of living in the countries of the North, declining long-haul travel costs, increasing holiday entitlements, changing demographics and strong consumer demand for exotic international travel have resulted in significant tourism growth to developing countries. Tourism is the principal export for one-third of developing countries. Tourism brings relatively powerful consumers to Southern countries, potentially an important market for local entrepreneurs and an engine for local sustainable economic development. There is no reliable data on domestic tourism but it is growing rapidly in South America and in China and South East Asia: it represents a very significant economic opportunity for many local communities.

Tourism and Aid

Multilateral and bilateral aid agencies are wary of involving themselves in the tourism sector. In 1969 the World Bank created a Tourism Projects Department recognising that in the Mediterranean and Adriatic countries, and in Mexico, tourism had been a significant generator of foreign exchange and of direct and indirect employment, internationally in the late nineteen sixties, there was considerable concern about high rates of unemployment and the ability of developing countries to service debt. Tourism sector studies were completed in some 31 countries and tourism staff regularly participated in World Bank macro-economic missions—their reports focussed on the potential for growth in tax revenues, foreign exchange earnings and direct and indirect employment effects. The primary emphasis was on national economic impact. By 1978 when the World Bank closed its Tourism Projects Department of the Bank had provided loans and credits for 18 projects in 14 countries and it was the major source of funds and technical assistance for tourism development. The bank withdrew from tourism development for a range of reasons amongst which were anxieties about the role of the bank in funding projects to develop luxury hotels designed to attract wealthy travellers from the developed countries. This strategy was seen inconsistent with new policy objectives which prioritised the bottom 40 per cent, the Bank's priorities were shifting towards the poor, a group, which was gaining relatively little from tourism development. There was a growing literature that focussed on the negative economic, social and cultural impacts of unmanaged tourism on local communities. The fuel crises of the nineteen seventies also undermined some of the forecasts that had been made for the strength of the market and the Bank withdrew from the sector in parallel with most other multilateral and bilateral agencies.

The international agencies followed a macro-economic tourism agenda in the nineteen seventies and eighties focussing on tax and foreign exchange revenues at the national level, major hotel and resort development, international promotion and national and regional master planning all attracted funding. In the nineties the adoption of the new poverty elimination target of halving the number of people living on less than 1 US $ per day by 2015 refocused development assistance on pro-poor growth. Multilateral and bilateral aid agency agendas are shifting towards micro economic growth strategies, which benefit local communities and in particular those below the poverty threshold. With poverty elimination now at the heart of decision aid, the potential for using tourism to generate pro-poor economic growth is being reassessed.

Since the mid-1980s, interest in 'green' tourism, eco-tourism and community tourism has grown rapidly among tour operators, policy makers, advocates and researchers. All of these focus on the need to ensure that tourism does not erode the environmental and cultural base on which it depends. The emphasis has been on minimising social, cultural and environmental impacts; rather than on positively affecting the livelihoods of the poor.

The Potential of Pro-Poor Tourism

There are a number of reasons to look again at tourism and to assess its potential to generate pro-poor growth. Eighty per cent of the world's poor live in just 12 countries and tourism is significant or growing in all but one of them. Tourism is a very large sector, it is growing rapidly, and there is some evidence that it is relatively labour intensive. The consumer travels to the destination, creating additional—local—opportunities for the sale of additional goods and services—ranging from local pottery to a guided walk. Tourism can be used to diversify local economies; it can often be developed in remote and marginal areas with few other diversifications or export opportunities. These areas often attract tourists because of their high landscape, cultural and wildlife values. These natural resources and the local culture are amongst the few assets of the poor.

Pro-poor tourism generates net benefits for the poor. It can be defined as forms of toursim where the benefits to the poor are greater than costs which tourism brings them. Economic costs and benefits are clearly important but social environmental and cultural costs and benefits are clearly important, but social benefits also need to be taken into account. Pro-poor tourism aims to expand opportunities for those living on less than 1 US$ per day. Whilst it will also need to be sustainable preserving local culture, minimizing environmental impacts, it will be driven by the poverty agenda. Community-based tourism seeks to promote initiatives by local communities or individuals within them; much has been learnt from these projects. Maximizing the poverty elimination effect requires that the emphasis is placed on involving those people who are living on less than 1 US $ per day and creating economic opportunities for them. Not all community tourism is pro-poor in this sense.

Effects on the Livelihoods of the Poor

Assessing the livelihood impacts of tourism is not simply a matter of counting jobs or wage income. Participatory poverty assessments demonstrate great variety in the priorities of the poor and factors affecting livelihood security and sustainability. Tourism can affect many of these, positively and negatively, often indirectly. It is important to assess these impacts and their distribution.

Tourism can generate four different types of local cash income generally involving different categories of people:

- wages from formal employement;
- earnings from selling goods, services, or casual labour (eg., food, crafts, building materials, guide services);
- and profits arising from locally owned enterprises
- Income: this may include profits from a community run enterprise, dividends from a private sector partnership and land rental paid by an investor.

Waged employment can be sufficient to lift a household from insecure to secure. But it may only be available to a minority, and not to the poor. Casual earnings per person may be very small, but much more widely spread and may be enough, for instance, to cover school fees for one or more children. Work as a tourist guide although causal, is often of high status and relatively well paid. There are relatively few examples of successful and sustainable collective income from tourism.

Negative economic impacts include inflation, dominance by outsiders in land markets and in-migration, which erodes economic opportunities for the local poor. Impacts differ between men and women. Women can be the first to suffer from loss of natural resources (e.g., access to fuel wood) and cultural/sexual exploitation, but may benefit most from physical infrastructure improvements (e.g. piped water or a grinding mill) where this is a by-product of tourism.

Positive Development Impacts of Tourism

On the positive side, tourism can generate funds for investment in health, education and other assets, provide infrastructure, stimulate development of social capital, strengthen sustainable management of natural resources, and create a demand for improved assets (especially education). On the negative side, tourism can reduce local access to natural resources draw heavily upon local infrastructure, and disrupt social networks.

Tourism affects the livelihoods of the poor by changing their access to assets. In several cases, tourism's impact on people's access to natural resources or physical infrastructure has been identified as the most important benefit or concern.

Cultural Impacts of Tourism can be Positive or Negative

Local residents often highlight the way tourism affects other livelihood goals-whether positively or negatively—such as cultural pride, a sense of control, good health, and reduced vulnerability. Socio-cultural intrusion by tourists is often cited as a negative impact. Certainly sexual exploitation particularly affects the poorest women, girls and young men. The poor themselves may view other types of cultural change as positive. Tourism can also increase the value attributed to minority cultures by national policy-makes. Overall, the cultural impacts of tourism are hard to disentangle from wider processes of development.

The overall balance of positive and negative livelihood impacts will vary enormously between situations, among people and over time, and particularly in the extent to which local priorities are able to influence the planning process. The application of a 'sustainable' livelihood framework is essential to developing pro-poor approaches. The distribution of livelihood impacts has to be considered. The poor are far from being a homogeneous group. The positive and negative impacts of tourism will inevitably be distributed unevenly among poor groups, reflecting different patterns of assets, activities, opportunities and choices. The most substantial benefits, particularly jobs, may be concentrated among few. Net benefits are likely to be smallest, or negative, for the poorest.

Policies to Enhance Pro-poor Tourism

Despite innumerable case studies of tourism development, there is relatively little assessment of practical experience in strategies to make tourism more pro-poor. Nevertheless, lessons can be drawn from a wealth of small initiatives (many from 'community tourism' or 'conservation and development' programmes), supplemented by expanding knowledge on 'pro-poor growth strategies', several policy implications clearly emerge.

1. *Put Poverty Issues on the Tourism Agenda*

A first step is to recognise that enhancing the poverty impacts of tourism is different from commercial, environmental or ethical concerns. PPT can be incorporated as an additional objective, but this requires pro-active and strategic intervention. There may well be trade-offs to make—for example between attracting all-inclusive operators and maximising informal sector opportunities, or between faster growth through outside investment, and slower growth building on local capacity. These trade-offs need to be addressed.

2. *Enhance Economic Opportunities and a Wide Range of Impacts*

Two approaches need to be combined:

- Expand poor people's economic participation by addressing the barriers they face, and maximising a wide range of employment, self-employment and informal sector opportunities;
- Incorporate wider concerns of the poor into decision-making. Reducing competition for natural resources, minimising trade-offs with other livelihood activities, using tourism to create physical infrastructure that benefits the poor and addressing cultural disruption will often be particularly important.

3. *A Multi-level Approach*

Pro-poor interventions can and should be taken at three different levels:

- this is where pro-active practical partnerships can be developed between operators, residents, NGOs and local authorities, to maximise benefits;
- national policy level—policy reform may be needed on a range of tourism issues (planning, licensing, training)

and non-tourism issues (land tenure, business incentives, infrastructure, land-use planning);

- International level—to encourage responsible consumer and business behaviour, and to enhance commercial codes of conduct.

4. *Work through Partnerships, including Business and Tourists*

National and local governments, private enterprises, industry associations, NGOs, community organisations, consumers, and donors all have a role to play. It is particularly important to engage business, and to ensure that initiatives are commercially realistic and integrated into mainstream operations. Private operators will not be able to devote substantial time and resources to developing pro-poor actions. NGOs and donors can help in reducing the transaction costs of changing commercial practice—for example, facilitating the training, organisation, and communication that would enable businesses to use more local suppliers. Changing the attitudes of tourists (at both international and national levels) is also essential if pro-poor tourism is to be commercially viable and sustainable.

5. *Incorporate pro-poor Tourism Approaches into Mainstream Tourism*

Pro-poor tourism should not just be pursued in niche markets (such as eco-tourism or community tourism). It is even more important that mass tourism is developed in ways that benefit the poor. It is also important to assess which tourism segments are particularly relevant to the poor. Domestic tourists are likely to be important customers.

6. *Reform Decision-making Systems*

It is impossible to prescribe exactly how each tourism enterprise should develop in ways that best fit with livelihoods. The most important principle is to enhance the participation of the poor. Three different ways of doing this can be identified:

- Strengthen rights at local level (e.g, tenure over tourism assets), so that local people have market power and make their own decisions over developments.

- Develop more participatory planning.
- Use planning gain and other incentives to encourage private investors to enhance local benefits. These approaches require implementation capacity among governmental and non-governmental institutions within the destination, and require a supportive national policy framework.

It is time to reconsider the role of tourism in contributing to pro-poor development. Tourism should be judged against other possible strategies and where it offers the best opportunities for pro-poor growth, or where it can make a useful contribution by increasing the diversity of opportunities for the poor, tourism it should be considered. However, careful and effective local management will be essential if it is to contribute to meeting poverty targets and if tourism dependency is to be avoided.

34
Consuming the Future

Now that we are to reach six billion of us, it is a good point to check again on what sort of lifestyles we pursue and what is the environmental impact of those lifestyles. It is curious that we have spent several decades being concerned about the growing numbers of humankind while not giving at least an equal amount of attention to the levels of living we aspire to, and how many natural resources we chew up thereby and how much pollution and waste we cause.

Everybody is a consumer of sorts. True, every fifth person scarcely qualifies for that designation, consuming goods worth less than $1 per day. Conversely, every seventh person qualifies for a designation of super-consumer, with a cash income at least fifty times greater. These latter are the people who, through their carbon dioxide emissions, are disrupting everybody's climate dozens of times more than the average citizen of One Earth. Fair play, anyone?

Much as the have-nots seek to match the have's, it is plain their efforts will not work out for a long time to come, at best. If every Chinese person were to consume just one additional chicken per year and if the said chicken were to be raised primarily on grain, this would account for as much gain per year as all the grain exports of the number two exporter, Canada. If the Chinese were to raise their per-capita consumption of beef, now only 4 kgs per year, to that of Americans, 45 kg, and if the additional beef were produced largely in feedlots after the manner of the United States, it would account for as much extra gain as the entire US grain harvest, less than one-third of which is exported. Because of its recent climbing up the food chain toward a meat-based diet, China has become one of the world's leading importers of grain. The global grain market today is around 200 million tons per year, and shows scant scope for significant increase.

As a further measure of its ambitions, the Chinese government has designated the auto industry as one of five industry "pillars". Today China has fewer cars than Los Angeles. If per-capita car ownership, together with oil consumption, were to match that of the United States, China would need 80 million barrels of oil per day—by contrast with the world's 1996 oil output of 64 million barrels of oil per day. The surge in carbon dioxide emissions would be unprecedented.

All this notwithstanding, there are already some 250 million newly affluent people in China. They are people with a household income equivalent to perhaps US$20,000, and enough discretionary income to enjoy the perquisites of the good life as perceived by these nouveaux riches. Top of the shopping lists are meat and more meat, followed by cars whether big or small. These are the badges of success: they show you have arrived.

The new consumers in China are matched by at least 200 million in India, and tens of millions in South Korea, Taiwan, Malaysia and Thailand (the recent economic setbacks have not permanently punctured the economic bubbles). Then there are 200 million more in Brazil, Argentina, Venezuela and Mexico, and more again in Hungary and other countries of Eastern Europe, also Turkey. Put them all together and they total about as many as the 800 million long established consumers in the ultra rich countries (the OECD grouping). When the current economic hiccups in Asia are left behind, the ranks of the new consumers can be expected to rise rapidly.

But they cannot hope to become super consumers. Where would all the extra gain come from? How could the global climate tolerate the huge additional pulse of carbon dioxide? There are all kings of other environmental reasons to suppose that environmental constraints will become all the more constraining. True, technology could help moderate the environmental impact. We could enjoy twice as much material prosperity while using only half as much natural resources and causing half as much pollution and waste. But the new consumers will want to pursue the American dream to the hilt, and it is hard to see that the best technologies could enable huge numbers of affluent aspirants, perhaps two billion people by 2010, enjoying even half the material prosperity of Americans with average household incomes of $40,000.

But it is true "prosperity"—mental and emotional as well as material? Or is the American dream becoming a nightmare with its harried lifestyles and declining leisure time, where the shopping mall is the ultimate Mecca, and the good life is a case of piling up goodies?

In any case, we cannot expect the new consumers to forgo their "rightful share" of affluence unless the longtime affluent agree to cut back on their environmental ruinous lifestyles. It is these communities that must offer a strong example, and soonest. Where is the political leader who will espouse the new vision, however much it may be perceived as the ultimate vote loser?

(35)

Genetic Diversity and Food Security

Mintaining a diversity of crops and varieties is a key to survival for millions of farmers living on impoverished land. For thousands of years, farmers have used the genetic variation in wild and cultivated plants to develop their crops and raise new breeds of live stock. Genetic diversity gives species the ability to adapt to changing environments, including new pests and diseases and new climatic conditions. Plant genetic resources—that component of genetic diversity of actual or potential use to humanity—provide the raw material for breeding new varieties of crops. These, in turn, provide a basis for more productive and resilient production systems that are better able to cope with such stresses as drought or overgrazing and can reduce the potential for soil erosion. The use of genetic diversity—on-farm, through field experimentation or in sophisticated gene transfer procedures—remains arguably the best route so securing our food and that of our children.

Although science has made enormous strides in improving the world's ability to feed itself over the past three decades, we cannot afford to rest idle. Nearly 800 million people in the developing world do not have enough to eat. In these regions, the rural poor represent about 73 per cent of the people living in poverty. They often live in marginal or unsuitable farming areas, such as zones with saline soils, and conditions, or degraded or hilly areas. Often isolated from other farms and far from urban areas, many poor farmers have barely benefited from agricultural developments elsewhere. In many cases they do not have access to commercially breed high yielding crop varieties. Diversity flourishes and remains important under such conditions.

Selections and Breeding

Poor farmers are well aware of the relationship between the stability and sustainability of crops and crop varieties on their lands. Their management and use of a diverse range of plants has often helped them to survive under the most difficult conditions. By growing a range of different crops, farmers have a better chance of meeting their needs. These might be crops that mature at different times or that can be easily stored to help to ensure a stable food supply throughout the year. They may also help farmers to provide a nutritionally balanced diet for their families, exploit different environment niches that exist on their land, or diversify their income sources.

Importantly, the genetic diversity contained in different varieties provides farmers with options to develop, through selection and breeding, new and more productive crops that are resistant to pests and diseases. The result may be a vast range of local varieties of crops grown by farmers in any one area.

Not respecting diversity can incur high costs: in 18th century Ireland, where potatoes were the only significant source of food for about one-third of the population, farmers came to rely almost entirely on one very fertile and productive variety, which proved susceptible to the devastating potato blight fungus. The resulting famine caused the death or emigration of more than 20 per cent of the population.

The value of diversity goes well beyond its ability to support stable production systems in marginal environments. As the world's human population rises, environmental problems (desertification, deforestation, erosion etc.) are intensifying. Climate change, particularly global warming, could bring about drastic changes in the location of the world's agro-ecological zones. Farmers will require new crop varieties capable of producing under diverse conditions, without adding ever-increasing amounts of fertilizers and other agro-chemicals. Because of the limited scope for growth in the world's cultivated areas, each new generation of varieties will have to be more productive than its predecessors.

Much has been written about the use of genetic engineering in plant breeding. Modern molecular techniques can be used to transfer genes from one living organism to anther or to change the genetic material within to produce more desirable traits. Genetic engineering has enormous potential to help solve problems that have proved intractable using conventional breeding approaches, such as developing crop varieties with in-built resistance to key pests and diseases and tolerance to stresses such as drought. However, the possible impact of these techniques, particularly on human health and the environment, is giving rise to fierce worldwide debate.

Take the case of banana and its close relative plantain, two of the developing world's most important crops. Their improvement is hindered by the sterility of most cultivars, a problem that can be addressed through genetic engineering. It is now possible to transfer gene constructs, such as those associated with disease resistance, directly into varieties with other desirable characteristics, drastically reducing the need for pesticides.

Today, research on genetic engineering is focussed on the development of commercial varieties of the world's major crops of interest to industrialized farmers. Many of the staple crops of importance to poor farmers in developing countries, such as cassava, bananas, beans and yams, have received relatively little attention. This situation is likely to continue as plant breeding is increasingly privatized and biotechnology becomes the fast-growing province of private industry. Meanwhile, the high costs of the new technologies are quickly exceeding the capacity of many, if not most, public research institutions—both in developing and developed countries—to support them. Thus, for the time being, increasing agriculture's role in the development of the world's poor is likely to continue to depend on the identification, maintenance and use of genetic diversity.

36

Economics and Sustainable Development

Economists and ecologists were once seen as enemies: environmental protection, it was thought, could only be achieved at the expense of economic growth. The misconception persists at the extremes among both the most fundamentalist Greens and the most ideological free marketers. But increasingly it is now being recognized that development and care for the environment go hand in hand. This interdependence is coalescing in the new and necessary discipline of environmental economics.

Conventional economic patterns have often assumed that growth and technical progress will nullify all resource and environmental limits. Environmental economics recognizes that the world's natural capital underpins all development, and that it is rapidly becoming scarcer as human demands exceed the globe's long-term carrying capacity. Government of India has introduced environmental measures over the last two decades, but need to move further towards integrating them into economic policies. There can be no real sustainable development unless environment and development policies are integrated at the very beginning of the decision-making process.

Quantifying the Environmental Cost

One of the first steps is to work out the true costs of polluting and depleting the world's natural resources, such as its soil, air and water, the climate and the ozone layer. These have often been regarded as free goods, and it was believed that the world has an infinite capacity to absorb the effects of human activities. Environmental economists, recognizing that the social and economic costs of degradation are very great, are trying to quantify them. They say this will make possible better use of such tools as cost-benefit analysis, environmental impact assessment and risk assessment—and the production of national income accounts which reflect the depletion and degradation of natural resources. As these costs are identified and quantified, economic policy can increasingly be developed with sustainable development as the primary objective. Achieving sustainable development requires industrialized and developing countries to make dramatic changes in national and international policies based on a global partnership. The greenhouse effect, the destruction of the ozone layer, the extinction of species and contamination of the oceans, and other environmental problems, affect us all, no matter which corner of the globe we inhabit.

The first and essential step in overcoming a difficulty is to recognize it and understand it. Concern over the difficulties related to sustainability has led scientists and national and international institutions to study the concept and suggest ways of meeting its many requirements. Indicators have been established to measure pollution levels, soil erosion, salinization, deforestation and a host of environmental problems. Evaluating the impact of such natural resource-use on ecosystems is a major step towards finding the necessary solutions.

For example, it has become clear, on a macroeconomic level, that national accounting systems fail to reflect these effects adequately. Deterioration of the world's rivers, land degradation, air pollution and contamination of the seas are not taken into consideration Inadequate accounting distorts reality and gives a false idea of the true consequences of growth and production.

On a microeconomic level, much is being done to redefine production costs. Incorporating the cost of waste management and internalizing negative external impacts within production prices are beneficial aspects of the economics of sustainability.

Steps are being taken to evaluate public and commonly held assets and to put a price on them, even though they may not be subject to market forces. These are only in the earliest stage but they will allow for more accurate evaluation of the world's natural capital. Fiscal, market, quota and other instruments are being developed to enforce change in the way in which certain resources are used. Examples include markets for transferable emission quotas or compensatory taxation mechanisms designed to ensure that economic forces act to reduce greenhouse gas emissions. Efforts at impact analysis—and in a general sense, cost-benefit analysis—permit rough estimations of the impact that projects might have on ecosystems.

Long-term Repercussions

These instruments carry significant limitations but they are important nevertheless because they attempt to quantify impacts on the natural world and to achieve a more rational use of natural resources. The development of such instruments and evaluation techniques will have significant repercussions in the formulation of sustainable long-term policies. But we must bear in mind that sustainability is not just an economic issue: it is also a political and cultural one.

The concept of sustainability demands as alternative view point in which humankind and the natural world are perception as a unit—as different yet mutually sustaining aspects of a whole. This perception is not incompatible with progress. It does not renounce development. It simply seeks to affirm life and refuses to discriminate between the means and the end. It understands that happiness cannot be achieved by destructive means. The questions of how to produce and how to consume therefore become extremely important. Neither should be at the expense of the future or of the natural world. Efficiency is not limited to the links between investment, products and prices: it must address the rational use of resources, including environmental and cultural consequences, both in the long and the short-term.

Very considerable adjustments must be made in the interests of sustainable development. They demand a reassessment of all our activities which cannot, logically, be done overnight. It is a long and continuous process, characterized by steadfastness and compromise.

37

Can Genetically Modified Organisms Feed the World?

The controversy over bio-technologies is raging. Advocates claim they're the only answer to malnutrition, while opponents warn that drought-resistant millet and vaccinated yams will only increase poverty. Farmers in arid and semi-arid regions of the world are anxiously waiting for rain to fall before they sow millet or sorghum, then hoe, harvest, feed their families and replenish their granaries. Meanwhile, researchers in Japanese, Chinese, Philippine, European and U.S. laboratories are making strides in sequencing the 12 chromosomes and 50,000 genes composing rice, the matrix of all grains and a staple for three billion human beings. In five to ten years, they hope to know enough to genetically modify not only rice, but millet, sorghum, manioc and sugarcane as well. The aim is to make them "naturally" resistant to drought, soil salinity, viruses, blights and other scourges.

Will these genetically modified organisms (GMOs) really guarantee "food security" in the short-term for the world's 826 million undernourished individuals? Will they help the small-scale farmers cultivating the Niger's barren, powdery soil to feed their families? In 50 years, the Earth's population will have soared to nine billion three billion more than today. And most of the newcomers will increase the already overwhelming pressure on the southern countries' much-depleted soil. The alarm has already been sounded for sub-Saharan Africa where, unlike India and China, the population growth rate is still sky-high and the number of undernourished people is barely declining. GMO supporters say only a major, revolutionary "technological leap" will enable the planet to feed all its children.

Experimenting with Miracle Seeds

But others strongly disagree, arguing that low food production is not what causes malnutrition. There is enough to eat in the southern countries, they say. But the world's poorest people, those with neither money nor land, living in disintegrating, war-torn countries, simply have no access to food. They argue that land-use conditions must change, poor people must have access to credit and local markets and small landowners must be freed from money-lenders. Better use could be made of traditional seeds instead of importing high-risk technology with unpredictable consequences, most of whose patents belong to giant multinationals.

Advocates of the GMO revolution work in bio-genetics laboratories, multinational seed and agrochemical companies, genome research, American foundations and some UN agencies, while most skeptics are out in the field. Adrao researchers have experience with miracle seeds. With international funding, they have developed a revolutionary variety of rice called Nerica. Genetically unmodified, it is the result of a conventional cross between a high-yield but fragile variety of Asian rice and a local variety that has had 35 centuries to adapt to Africa's stressful environment. Nerica offers tremendous possibilities. It reaches maturity in 90 days instead of the usual 120 to 150, resists insects, yields three tonnes per hectare with neither fertilizers not irrigation—compared with 1.5 tonnes for traditonal varieties—and grows like a weed. Ideally, it should improve life for hundreds of thousands of small-scale farmers who practice pluvial rice farming on plots of land ranging in size from 20 to 200 square metres and help West Africa's countries drastically cut their rice imports, perhaps even to export the grain.

Fear and Red Tape

However, this breakthrough is having difficulty leaving the laboratory. Red tape, administrative delays and lack of communication between ministries and farmers, seed-certification organizations and rural credit institutions are to blame. A "technological leap" is not likely to help matters much.

Several African, Asian and South American countries have already passed laws regulating GMO production. But can they enforce them? Which laboratories will monitor the changes in bio-diversity that might result from unforeseen crossbreeding between GMOs and related wild species? And with what funding? Who will see to it that pollen from GMOs capable of transmitting their defense mechanisms against insects and viruses do not spread? Researchers retort that it would be a mistake to focus on first-generation GMOs, which are necessarily flawed. "Soon we'll see the appearance of the second, third and fourth generations, which will meet developing countries' needs better".

Bio-genetics makes it possible, for example, to insert an insect's gene into a plant or vaccines into bananas or potatoes. To create a variety of rice that needs no more water than a camel (instead of the 4,000 to 5,000 litres necessary to produce one kilo), or again, to enrich plants with vitamins and minerals and develop others that revitalize acidic soil devastated by over-farming. So why not carry out the wildest projects?

It is easy to understand the enthusiasm of certain researchers and philanthropic institutions. For example, the Rockefeller Foundation sees what it calls the biotech-driven "second Green Revolution" as a way to make up for the first one's mistakes and tragedies. True, in the 1960s, the Green Revolution helped double food production by creating high-yield varieties of wheat and rice, keeping pace with the world's population growth. But these seeds, which need plentiful inputs (irrigation, fertilizers, herbicides and pesticides), have primarily benefited farmers who could afford to invest. Africa and the poorest areas in Asia and Latin America were left out. Moreover, the results for beneficiaries, such as China and Vietnam, are mixed: traditional varieties have vanished, irrigation has increased soil salinity, and farmers have over-used herbicides and pesticides to the detriment of their health and the environment.

Great Expectations

Supporters of the "second Green Revolution" say GMOs should spark an explosive increase in yields without inputs and in extreme farming conditions. But will they benefit the poorest farmers? Until now, multinational agro-chemical companies converted to the "life sciences" have focused all their investments on intensive crops with close connections to industry and built a will of prohibitively-expensive patents around their discoveries. Only publicly-funded research has taken an interest in indigent farmers living in tropical areas. Strapped for cash, the public sector has been forced to sign cooperation agreements with the private sector at the risk of losing its independence.

Major biotech companies are quickly caught onto the image enhancing advantages of helping to develop GMOs for the Third World. In 2000, amid a blaze of publicity, biotech heavyweights granted the free use of 70 patents to help develop a genetically-modified variety of rice. The grain was heralded as a "miracle rice" capable of conquering Vitamin A deficiency, which kills one to two million children each year. But so far, the "golden rice" has fallen far short of expectations. The publicly-funded, Philippines-based International Rice Research Institute (IRRI) estimates that it will take five to ten years before seeds can be distributed free to farmers with annual incomes of under $10,000, in line with the agreements signed with industry. Furthermore, it is still unclear how much of the rice has to be consumed to make up for a Vitamin A deficiency.

For non-governmental organizations campaigning to protect the environment and preserve bio-diversity, such as the Rural Advancement Foundation International (RAFI) network, this deal "could kneecap other, low-tech and more cost-effective solutions, such as re-introducing the many vitamin-rich food plants that were once cheap and available."

Lessons from the Past

Will GMOs help wipe out malnutrition? The golden rice episode has set the tone for the debate. Advocates say it would be utopian to wait for a better world while existing technology can help solve the problem here and now. "Some argue that lack of food is simply a problem of unequal distribution. If poor people were not poor, they could buy the food they need, "This is true, but over-simplistic. There are no signs the world is about to engage in a massive redistribution of wealth." First it is necessary to improve the conditions of agricultural production and soil management, keep the ground from getting hard after clearing and decrease rice imports by impoverished West African countries. All that can be achieved without GMOs, which run the risk of impoverishing bio-diversity."

38

Crisis Prevention

Can Better Development Planning Lessen the Toll of Civil Emergencies and Natural Disasters?

Even a cursory scan of the world's headlines is depressing: armed conflicts are grinding on in Somalia, Afghanistan and in a growing number of other countries. And the effects of natural disasters are becoming more catastrophic each year. International relief aid, in response to such emergencies, has increased substantially. But how large can these sums of money realistically be expected to grow? With no end in sight to the need for relief, the goodwill of international donors is quickly giving way to disillusionment.

This leads us to a second question, which is, where does development fit in this grim scenario? For the development community to remain aloof from the issue of disasters and emergencies is not only politically short-sighted, it also ignores totally the causes and the effects of such phenomena.

Natural hazards such as hurricanes and earthquakes may be impossible to prevent. But they only become natural disasters if people are vulnerable. Why is it, for example, that an earthquake in Khilari, Maharashtra that registered 6.9 on the Richter scale killed up to 35,000 people, when an earthquake of almost the exact same magnitude in Los Angeles in 1994 claimed only 57 lives? By reducing poverty we can help increase the coping capacity of vulnerable populations. Therefore helping people lower such vulnerability is as much a development issue as the environment, or women's participation in development. Moreover, the repercussions of natural disasters go far beyond the immediate casualty list that so transfixes the media. Secondary and longer-term effects can be equally if not more devastating. And they must be taken into account by development practitioners.

It has been estimated, for example, that the damage to Mexico City's infrastructure from a massive 1985 earthquake amounted to US$ 3.6 billion. Yet over the subsequent five years, the negative ripple effect on that country's balance of payments resulted in a loss of $8.6 billion. Furthermore, reconstruction requirements forced Mexcian authorities to revise their economic polices to meet an increased demand for public funding, credits and imports. The priorities for public expenditure were redirected to reconstruction projects, leaving many of the pre-disaster problems of the city and its people unattended.

In Bangladesh, floods in the recent past claimed 2,000 people. But on closer examination we find that the toll was much more exensive than that: in each of these years the country's economic growth rate was halved by the delayed planting of rice and the destruction of seedbeds in the floods, further undermining the country's food security. All of these are considerations that go beyond relief, but they must be taken into account by development professionals.

Other emergencies may be more complex, but must be subjected to the same analysis. As the situations in Angola, Burundi, Somalia and the former Yugoslavia demonstrate, we know little about the dynamics of emergencies that arise from civil conflict. We do know, however, that their cause usually lies in a lethal mix of poverty, poor governance and ethnic or religious rivalries exacerbated by profound social inequities. We are also learning that their resolution frequetly requires the application of peacekeeping and political measures, combined with relief and development. Among the most virulent effects of such complex emergencies is the massive displacement of people; women and children are the principal victims, constituting 70 per cent of the world's refugees.

These complex emergencies around the world could easily get worse before they get better. This being said, carefully designed development efforts—carried out as building blocks to national reconciliation in the fragile post-conflict stage will need to increase commensurately. The appropriateness and the sustainability of these development efforts will be one of the most important factors in determining whether peace itself becomes sustainable. For example, the absence of carefully tailored reintegration strategies for demobilized soldiers and their host communities would be an almost open invitation to resumed violence.

Yet we must also be conscious of the impact of aid and try harder to prevent the need for relief in the first place. An increasing body of evidence suggests, for example, that emergency aid can sometimes be counter-productive in the longer term, increasing the vulnerability of populations and impeding recovery. Ironically, we find ourselves in situations today where it is far easier to obtain funds for maintaining refugees in their places of asylum than for helping them reintegrate into their own societies. In such cases, we may very well be helping to perpetuate the problem that we sought to relieve, as the presence of larger numbers of refugees is sometimes itself a cause of conflict.

So how are we to proceed? And what exactly is the nature of the relief to development continuum that remains logical in the abstract but elusive in reality? The concept of a continuum does not imply a linear and absolutely progressive set of responses. On the contrary, it means that we are dealing with a set of processes rather than rigidly defined steps. It also means that development must be very much part of the disaster management process, and that the aim of the continuum must be to move from relief to rehabilitation and resumed development at the earliest opportunity. However, this resumed development must include conscious measures to reduce the vlunerability that caused the disaster or the emergency in the first place.

In other words, we must give greater thought to prevention before we reach for the "cure"—for humanitarian, political and financial reasons. (The Japanese insurance industry spends $ 200 million a year on disaster education alone). And as development paractitioners, we must reconcile ourselves to the vastly more complicated environment in which we have to operate.

This means, for example, that we will have to begin examining whether the economic policy "medicine" often prescribed will reduce conflict or enhance it. We will have to ask ourselves if the reconstruction period following a civil conflict or natural disaster is the right time to advocate cuts in social spending, as has happened in certain countries in Africa and Latin America. Similarly, is it really in children's best interests to build a school in a seismic zone without first ensuring its structural stability? And does it really make sense to urge drought-prone countries to increase their reliance on cash crops, as has been done in some instances.

A story that never made headlines anywhere involves hundreds of the poorest people in Bangladesh, whose homes remained intact during the floods of 1988, when many others were simply washed any. These people were fortunate enough to have obtained credit through the Grameen Bank for construction materials as well as instruction in the building of flood-resistant homes. The Grameen revolving fund had received start-up capital from international financial agencies. Since that time the effort has been expended, and more than 10,500 flood-resistant homes have been built in the last two years.

This is just one example of the kind of action we need more of in fairly predictable and recurring circumstances such as the floods in Bangladesh, as well as in the more complex, man-made emergencies to which we must respond.

Bibliography

Anand, R.P., *Legal Regime of Sea Bed and the Developing Countries*, 1975.

Bhatt, S., *Environment Protection and International Law*, Radiant Publication, New Delhi, 1985, p. 122.

Bhatt, S., *Environment Laws and Water Resources Management*, Radiant Publication, India, and Advent Books Inc. New York, 1986, p. 355.

Behrman, Danial, *In Partnership with Nature—UNESCO and the Environment* (Paris, 1973).

Bell, Daniel, "Technology, Nature and Society", *American Scholar*, Summer 1973.

Bentley, Glass, "Biology and Human Values", USIS, New Delhi.

Book of Nature: The Way Things Work, Published by George Allen and Unwin Ltd., 1981, p. 525.

Boulding, Kenneth E., "New Goals for Society", S.H. Schun, ed., *Energy, Economic Growth and the Environment*.

Carr, E.H., *What is History*, 1961.

Darlington, C.D., *The Evolution of Man and Society* (London, 1961).

Downing, Paul B., ed., *Air Pollution and Social Sciences* (New York, 1971).

"Drive to adopt national water policy", *Times of India*, 22 July, 1983.

Dubos, Rene, "Man and his Environment", *Britannica Perspectives*, Vol. 1, 1968.

Einstein, A., *My Views*, ed. By S.K. Bandopadhyaya (Calcutta, 1976).

"Environment Research Programme", prepared by NCEPC, Department of Science and Technology, New Delhi.

Forbes, R.J., "The Conquest of Nature and its Consequences", *Britannica Perspectives*, Vol. 1, 1968.

Fowler, John M., *Energy and Environment* (New York, 1975).

Fuller, Buckminister R., *Operating Manual for Spaceship Earth* (New York., 1969).

Gandhi, Indira, "Poverty Greatest Pollution, says Mrs Gandhi", *Times of India*, 8 September, 1981.

Glenn, Seaborg, " Science, Technology and Development: A New World Outlook", USIS, New Delhi.

Hacoley, Amos H., *Human Ecology* (New York, 1950).

"India must develop own ecology", *Times of India*, 8 October. 1981.

Marion, Jerry B., *Energy in Perspective* (London, 1974).

Misra, K.C., *Manual of Plant Ecology*, New Delhi, 1980.

Mukherji, P,K., *Life of Tagore*, trans, by S.K. Ghosh, 1975.

Mumford, Lewis, "The Future of Cities", in *Basic Issues in Environment*, E.J. Winn, ed., 1972.

Palmslierna, H., *Future Imperatives for Human Environment*, 1972.

Pavithran, A.K., "World Futurology", *Eastern Journal of International Law* (Madras), vol. 9.

"Plans to usher India into 21st Centruy", *Times of India*, 24 October, 1985.

Polunin, Nicholas, "The Biosphere Today", *The Environmental Future*, Proceedings of Ist International Conference on Environmental Future in Finland, ed. By N. Polunin, 1972.

Radhakrishnan, S., *Recovery of Faith, 1967.*

Report on the State of Environment prepared by Centre for Science and Environment, New Delhi, 1985.

Sarkar, Mahendra Nath, *The Cultural Heritage of India*, Vol. 1.

Sen, Sudhir, "Blueprint for a Better World", *Times of India*, 2 March, 1980.

The Limits to Growth, a report to Club of Rome (New York, 1972).

The Mind of J. Krishnamurti, ed. by L.S.R. Vas (Bombay, 1971).

Toynbee, Arnold, "an and his Soul", *Hindustan Times*, 4 January, 1968.

United Nations List of National Parks and Protected Areas, 1985.

Vivekananda, Swami, *Complete Works*, vol II (Calcutta, 1968).

Ward, Barbara and Dubos Rene, *Only one Earth: The Care and Maintenance of a Small Planet*, Report to UN Conference on human environment, Stockholm, 1972.

"Wildlife laws in India", *Times of India*, 4 March, 1985.

Word, Barbara, *Progress for a Small Planet*, 1979.

Index

U

W